Developing Number Sense

Also available from Continuum

Teaching Number Sense 2nd Edition, Julia Anghileri

Inclusive Mathematics 5–11, Brian Robbins

Teaching Mathematics in the Primary School, Gill Bottle

Getting the Buggers to Add Up 2nd Edition, Mike Ollerton

100 + Ideas for Teaching Mathematics, Mike Ollerton

Developing Number Sense

Progression in the Middle Years

Julia Anghileri

Continuum International Publishing Group
The Tower Building
11 York Road
London
SE1 7NX
www.continuumbooks.com

80 Maiden Lane, Suite 704
New York, NY 10038

British Library Cataloguing-in-Publication Data
A catalogue record for this book is available from the British Library.

ISBN: 9781847061263 (paperback)
9781847061256 (hardcover)

Library of Congress Cataloging-in-Publication Data
Anghileri, Julia.
Developing number sense : progression in the middle years / Julia Anghileri.
p. cm.
Includes bibliographical references and index.
ISBN-13: 978-1-84706-125-6 (hardcover)
ISBN-10: 1-84706-125-7 (hardcover)
ISBN-13: 978-1-84706-126-3 (pbk.)
ISBN-10: 1-84706-126-5 (pbk.)
1. Number concept in children–Study and teaching (Elementary) 2. Mathematics–Study and teaching (Elementary) I. Title.

QA141.15.A54 2008
372.7–dc22

Typeset by YHT Ltd, London
Printed and bound in Great Britain by MPG Books Ltd, Bodmin, Cornwall

To Bob, my constant mentor, and to Nikki, Dominic, Ben and Tom who have each taught me so much.

Contents

List of Figures

CHAPTER 1

Numeracy in the Twenty-first Century

There have been many changes in mathematics teaching over the last few decades and children in the mathematics classroom today will be involved in many more thinking activities and much less 'drill and practice'. The expectations for children's achievement have changed with the introduction of new technologies, such as computers and calculators, and with topics such as data handling established alongside more traditional aspects of mathematics. There have been fundamental changes to the curriculum with the word 'numeracy' introduced to convey a meaning of not only proficiency with numbers, but confidence and inclination to use numbers in practical problem solving, in familiar and in novel contexts.

In number work, what is needed today is that children make sense of numbers and approach problems with a repertoire of strategies for calculating and not just the standard methods. This expectation of flexibility is illustrated well in the types of question that are included in national assessment tests. For example, in 2003 at Key Stage 3 question 3 on paper 1 asked

b) How many sixths in $3\frac{1}{3}$ [only a third of pupils were successful]

c) Work out $3\frac{1}{3} \div \frac{5}{6}$

The comment in the report stated that even among those most able pupils who were successful in part b) 'there was very little evidence of pupils reasoning that they simply needed to divide their answer for part b) by five' (QCA, 2004b: 20).

In Key Stages 1 and 2 there is also this expectation for children to look for ways to make sense of questions, and many assessment questions are designed to elicit thoughtful strategies rather than the standard approaches that may have been appropriate for previous generations.

Background to change

Theories of learning

One reason for such changes in the teaching of mathematics is that theories of learning have developed from the 'behaviourist' theories that influenced teaching in the 1930s and 1940s, to the 'constructivist' theories that are relevant today. Earlier last century, influenced by a behaviourist paradigm for learning, it was thought that children learned best by being trained, with a stimulus and reward system and plenty of practice. Exercises in arithmetic consisted of many sums to be solved using standard methods that were demonstrated by the teacher and practised by the pupils. Many algorithms had to be memorized and there was little room for deviating from these methods or for discussion. Some of the methods taught were hard to understand and many pupils found arithmetic frustrating and difficult. The result is that there are many negative feelings about mathematics among adults today and in calculating they often use methods that are different from the ones they were taught in school.

More recently, psychologists have shown that children learn more readily, and develop a better range of skills, if they are actively involved in constructing their own meanings. Within the constructivist paradigm for learning, 'coming to know' is an adaptive process that organizes one's experiential world, with knowledge actively constructed by the individual (see for example Jaworski, 1988). It is not just a case of memorizing those procedures that are taught, but constructing an understanding that enables individuals to develop new skills that will be transferable to a range of problems. Classrooms are a place for discussion and negotiation as different methods of calculating are compared for their efficiency and ease of understanding. The intention is to give every pupil the confidence to tackle problems by thinking through the mathematics involved and identifying an appropriate solution method. This has made life very different for teachers who are no longer there to demonstrate the standard methods for calculating but who now have the responsibility to promote the pupils' own thinking, encouraging efficiency without loss of understanding, helping each individual to work within their capabilities.

Recent research has shown that teachers of mathematics are more successful when they believe that the most important focus for teaching is that of making mathematical connections. Making connections, for example between 2×3 and 2×0.3 or 0.2×0.3 will enable children to make sense of many calculations by relating them to known results. It is not just a connection with known facts but also with experiences and imagery that will help them understand the processes involved so that they can remember them more easily, reconstruct them if they are forgotten, or adapt them for a new situation. This has been confirmed in research where the study of effective teachers of mathematics identified a *connectionist* approach as being more effective than a traditional *transmission* approach, where teachers specify particular methods to be used, or a *discovery* approach where children are left to find their own strategies (Askew et al., 1997).

Past traditions

In order to understand some of the most recent changes, it may be helpful to look at some extracts from a book on arithmetic written early in the last century. At this time, arithmetic teaching revolved around the algorithms for the 'four rules' of addition, subtraction, multiplication and division and pupils were expected to practise the standard methods by completing pages of sums. In Schonell's book for teachers, the author identifies a progression for teaching calculating and the following extract (Figure 1.1) concerning subtraction analyses the difficulties that arise at different stages (Schonell, 1937).

The vocabulary of 'subtrahend' and 'minuend' may be unfamiliar today but most adults will recall the idea of 'borrowing' although not all will understand why it works. Before reading on, look at step 14 and take a few minutes to decide how you would tackle the final calculation 6067 – 5970 by using an algorithm or by some more informal method. In today's curriculum we encourage children to think about an appropriate way to do each calculation while in 1937 it was made clear that a standard method was required and little emphasis was placed on understanding. The following quote (Figure 1.2) from the same text would make shocking reading for today's teachers.

1st step :

98	57	84	38
3	4	1	8

Tens and units in minuend; units in subtrahend; no borrowing.

2nd step :

55	99	78	97
32	43	10	22

Tens and units in minuend and subtrahend; no borrowing.

8th step :

316	564	68	387
27	59	59	299

Borrowing in units and tens or borrowing in units and zero result in tens.

9th step :

80	980	430	168
57	930	416	68

Introduction of zero difficulty in units or tens.

11th step :

346	629	756	387
284	473	382	196

Borrowing in tens place. Numbers over 100.

12th step :

364	831	8354	8112
295	276	5676	6798

Borrowing in hundreds, tens and units places.

13th step :

800	607	700	906
695	298	192	199

Advanced " 0 " difficulties and borrowing.

14th step :

891	904	705	6067
207	206	109	5970

Advanced " 0 " difficulties and borrowing.

Figure 1.1 Steps in subtraction identified by Schonell

BACKWARDNESS IN ARITHMETIC 73

(a) *Over-explanation of Processes with Duller Pupils.*—There are some pupils, mainly the independent enquiring ones of supernormal intelligence, who must have, and who thrive on, explanations in arithmetic, but there are few pupils in "B" or "C" classes who require much explanation. What they require to know is how to get the sums right, and when they have learnt the method so thoroughly that the possibilities of getting the particular type of sums wrong are only the ordinary ones due to chance, then explanations might be attempted. Although most adults can divide one fraction by another or calculate square root, it is doubtful if more than 25 per cent of them can explain their methods of working. Why then should we burden children with unnecessary explanation in arithmetic ?

Figure 1.2 A quote from 1937 about backwardness in arithmetic

Many of the older generation will remember the difficulty of trying to remember all the rules in arithmetic, and the frustration of being unable to reconstruct a method that was forgotten. Few adults will remember the method for finding square roots by calculating despite the fact that it was taught and practised by many. The emphasis today is on pupils developing efficiency in calculating without loss of understanding, and on sharing their understanding by explaining and listening to the explanations of others, and this has made the mathematics classroom a more comfortable place than it has been in the past. That does not mean that children are not challenged, but the demands are different as the emphasis is on developing flexible problem solvers rather than calculators who are disciplined in standard methods. The National Curriculum (www.nc.uk.net) identifies key roles for 'communicating' and 'reasoning' in mathematics learning and the National Numeracy Strategy Framework (DfEE, 1999) includes 'reasoning about numbers' and 'explaining' in the teaching programmes for every year group.

Changes to classroom practices

Changes in the classroom have been influenced by several government reports that acknowledge the importance of mathematics for society today and that reflect the growing understanding of how children learn (Anghileri, 2001). An early Committee of Enquiry Report, *Mathematics Counts* (Cockcroft, 1982), endorsed the emphasis on teaching mathematics for understanding and involving children actively in their own learning. Among the recommendations has been a shift from classrooms where teachers did all the talking, to engaging pupils in discussion with the teacher and with their peers. It is expected in classrooms today that teachers will encourage their pupils to reason and communicate their ideas, building on their intuitive thinking, so that they gain the confidence that will be needed for tackling mathematical problems. In addition to the three strands on number – 'counting and understanding number', 'knowing and using number facts' and 'calculating' – the Primary Framework (DfES, 2006a) includes a strand for 'using and applying mathematics' that identifies particular skills that will be the basis for mathematical thinking and communicating. For Key Stage 1 'explaining methods is an important foundation for reasoning and proof in later key stages' and 'making and using pictures, diagrams and notes to aid thinking are important skills'. As they progress through Key Stages 2 and 3 'communicating solutions, decisions, explanations and reasoning to others form another set of skills that children need to learn'. They need to plan their enquiry and learn to predict, hypothesize and test as they develop their mathematical understandings (DfES, 2006b). It is no longer sufficient to work in isolation at solving problems as society today relies on teams of workers who can communicate and support each other in collaborative thinking.

Recommendations in the National Numeracy Strategy Frameworks for Teaching Mathematics note that better standards of mathematics occur when 'regular oral and mental work develops and secures pupils' recall skills and mental strategies, and their visualisation, thinking and communication skills' (DfEE, 2001: 6). A structure for most mathematics teaching is recommended in the 'three part lesson' consisting of an 'oral and mental starter', 'a main activity' and a 'plenary'. This enables pupils to be engaged actively in their learning rather that being passive recipients of knowledge. The oral and mental starter focuses on the understanding that children need rapid recall of number facts and mental strategies for solving problems, although this must be balanced with the need for thinking time in developing strategies. Through careful questioning children will become aware that mathematics is not just a series of isolated bits of information to be recalled, such as the multiplication facts, but that everything connects in a meaningful way. Recall of facts is supported by mental strategies for deriving new facts from those that are already known. For example, the result of multiplying 6 × 7 can be related to multiplying 60 × 7, thus making a calculation like 61 × 7 accessible as a mental calculation, perhaps supported by a jotting. Through the main activity these connections can be reinforced in problem solving where groups may work together to utilize their individual understandings and collaborate in developing new strategies.

The plenary offers opportunities for teachers to value and share the pupils' own suggestions developing 'taken as shared' meanings and encouraging *conceptual discourse* (Anghileri, 2006a). By asking questions such as 'which calculations were the most difficult?' and 'what made them difficult?' they can raise children's awareness of significant factors. As the children verbalize and listen to different strategies the teacher can help them identify key characteristics of good mathematical thinking and begin to establish mathematical conventions. The plenary also presents opportunities for informal assessment as each child's contribution will indicate the conceptual level at which they are working and the efficiency of their solution strategies.

Central to working with numbers is the development of a whole range of mental strategies, and these are given priority as the National Curriculum for primary school states that 'pupils are expected to use mental methods if the calculations are suitable' (DfEE, 1999: 69). A level 3 question in the 2003 Key Stage 1 assessment was the calculation

$$176 - 49 = \square$$

but the report noted that 'very few children used the fact that 49 is almost 50' to make the question accessible mentally, rather they used more standard methods to subtract 40 and 9 (QCA, 2004a). Pupils are also expected to learn efficient written methods, but not at the expense of understanding, and these written methods will be developed progressively through their schooling. The changed emphasis for teaching is summarized in the curriculum for Scotland as follows:

- mental calculation will require more attention
- written methods need not always be standardized on a concise but difficult to understand algorithm – we do not need, for example, to set out numbers vertically for real life subtraction
- new skills in using a calculator will need to be learned
- pupils will require the ability to decide which method, mental, written or using a calculator, is appropriate in particular circumstances.

(Learning and Teaching Scotland, 1991)

In the Framework for Mathematics in England there is an expectation that 'written methods for addition and subtraction should be well established by Year 6 for nearly all pupils ... [but written] multiplication and division methods will need to be developed further' in Key Stage 3 (DfEE, 2001: 11).

Scaffolding children's thinking

The skill of making a decision about how a calculation may be done needs to be nurtured as many pupils still rely on their teachers to guide them to the most appropriate method. Children need to develop confidence in their own decision making and this may mean they spend some time working inefficiently as they try different methods. The challenge of tackling new problems, retaining ownership of solution strategies and making connections with existing knowledge are crucial experiences that will take time in the classroom so that teachers will need a range of approaches for supporting individuals. Three levels of scaffolding practices to support mathematical leaning are identified by Anghileri (2006a) from level one that includes *environmental provisions* and *classroom organization* through level two; interactions that are increasingly directed to developing richness in the support of mathematical learning through *explaining, reviewing* and *restructuring* to level three; where the focus is on *developing conceptual thinking*.

In analysing the way teachers interact with their pupils at level two subtle but important differences are identified as *reviewing* scaffolds and *restructuring* scaffolds:

> When students are engaged with a task, they are not always able to identify those aspects most pertinent to the implicit mathematical ideas or problem to be solved. A response for teachers is to refocus their attention and give them a further opportunity to develop their own understanding. *Reviewing* classifies five such types of interaction:
>
> - getting students to *look, touch and verbalise* what they see and think;
> - *interpreting students' actions* and talk;
> - using *prompting and probing* questions;
> - *parallel modelling* where a problem with similar structure is solved and
> - getting students to *explain and justify* what they have done.

In contrast,

> through *restructuring*, the teacher's intention is to introduce progressively modifications that will make ideas more accessible, not only establishing contact with students' existing understanding but taking meanings forward. This differs from *reviewing* where teacher-student interactions are intended to encourage reflection, clarifying but not altering students' existing understandings. Restructuring involves interactions such as:
>
> - provision of *meaningful contexts* to abstract situations;
> - *simplifying the problem* by constraining and limiting the degrees of freedom;
> - *rephrasing students' talk* and
> - *negotiating meanings*.
>
> (Anghileri, 2006a)

For level three teachers go beyond explanations and justifications to engage their pupils in *conceptual discourse* that extends their thinking by initiating reflective shifts so that what is said and done in action subsequently becomes an explicit topic of discussion. This involves metacognitive processes through which children reflect on what they have said and done, for example not only identifying effective solution strategies for problems, but going further to discuss *why* these methods are effective. Another situation enabling conceptual discourse would be to identify why some calculations are easy while others are more difficult. This will include talk about the way the different numbers can affect the strategy for a calculation, and why some word problems are difficult to convert into mathematical calculations.

Negotiating meanings

In acknowledgement that children are actively involved in constructing their own understandings, teachers have moved from 'telling' pupils what is needed, to listening to pupils, building on informal understandings, and negotiating new meanings for words and symbols as more mathematics is introduced (Anghileri, 1995). In helping to extend their pupils' existing knowledge, teachers will take account of the interpretation brought to a problem by the individuals, as well as the mathematical significance of any language that is involved. In the example earlier in this chapter, 6067 – 5970, it is helpful to interpret the expression as 'the difference between 6067 and 5970' as the idea of 'take away' may not fit this calculation so readily. In contrast to past practices where children were first taught how to do a calculation and then later applied this to a problem, many examples today start with numbers set in a context that will generate the need for a calculation procedure. Tackling context problems will be typified, in the early stages, by the close association between (even inseparability of) the words describing the real context and number operations for solving particular problems (Figure 1.3). Only later will the formal mathematical language and symbols be understood.

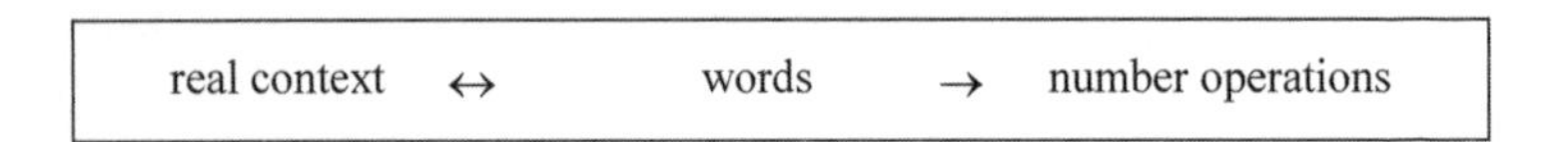

Figure 1.3 Associating words, contexts and number operations

The experiences children get in solving real problems will form the basis for later strategies and the contexts can provide important models and visual images for calculating strategies. For example, young children will need to physically 'take away' objects from a set to understand subtraction at a very limited level. At this stage, interpretations of the operations will be restricted by everyday language that will have limited meaning. Subtraction will later be related to the idea of 'difference' as in the calculation 'Jay is 12 years old and Jeva is 8 years old. What is the difference in their ages?' Here the image may be 12 years and 8 years displayed on the same time line so that the difference is physical. Both 'take away' and 'difference between' will be recorded succinctly with the '–' sign, which will come to have different meanings and lead to different solution strategies for problems.

Because the way in which symbols are read will influence the calculation method chosen, restrictions in interpreting symbols can have an inhibiting effect when calculations are introduced symbolically. It is often thought that mathematical terms and symbols are concise and unambiguous, but there are complexities with the language at every stage. In a discussion of words with multiple meanings, a classic example is given by Pimm (1987), who cites two pupils who respond to the question 'What is the difference between 24 and 9?' saying 'One's even and the other's odd' and 'One has two numbers in it and the other has one'. Here the pupils fail to understand that the term *difference* is being used in a mathematical sense.

Mathematical symbols provide a concise way of working and communicating, but different interpretations can confuse children, and even undermine their confidence. The symbol ' × ', for example, is commonly read as 'lots of', 'times' or 'multiplied by'. Each word or phrase indicates an interpretation that may be matched to a procedure for solving problems. Where '3 × 4' is interpreted as '3 times 4' or '3 lots of 4', this relates to 3 sets of 4 objects (4 + 4 + 4), while '3 multiplied by 4' would be 4 sets of 3 objects (3 + 3 + 3 + 3). This will make a difference when '3 × 20' is interpreted as '20 + 20 + 20' or '3 + 3 + 3 + . . . + 3' twenty times. Neither of these interpretations will be of much help with the problem '0.3 × 0.4' and the language for interpreting this problem will need to be modified to match this new situation.

In division the confusion that arises is perhaps more acute because this is generally the hardest operation for children to master as it builds on all their prior understanding of numbers. Tape recording children and teachers as they work on division tasks has provided evidence of wide variations in the phrases used to read and interpret the division symbol. By listening carefully to children the following phrases, all involving the word 'divide', were heard for the problem '12 ÷ 3':

12 divided by 3	12 divided into 3
12 divide by 3	12 divide into 3
12 divided into 3s	

Depending on the phrase chosen, different visual images of both the procedure for solution and the outcome could be different. The *passive* construction 'divided by' was replaced by the *active* construction 'divide by' to indicate a procedure that may be implemented to solve the problem. Just as a pizza may be 'divided into' three portions, some children appear to associate division with partitioning *into* equal subsets. There is a subtle difference implicit in the phrases '12 divided into 3' and '12 divided into 3s' where the first may be identified with a sharing procedure resulting in 3 'portions' while the second suggests a grouping procedure or repeated subtraction of 3s.

As well as these phrases using derivatives of the verb 'to divide' there appear to be a wealth of alternatives used by children:

shared into 3, shared into 3s, shared by 3, shared with 3
split into 3, split into 3s
how many 3s in 12, 3 into 12, 3s into 12
12 grouped in 3s, 12 grouped into 3s, 12 grouped in 3

It appears that teachers may 'accept' these phrases from the children while using the formal interpretation '12 divided by 3' themselves. Without awareness of the children's interpretation there is much scope for confusion (Anghileri, 1995).

In addition to this variety of phrases for interpreting division there are two distinct symbolic forms to represent the same operation as 28 ÷ 7 or as $7\overline{)28}$ (that are read as '28 divided by 7' and as '7 divided into 28'). Further ambiguity can arise when these symbols are read out and this possibly explains the plethora of interpretations for division identified above. On meeting the calculation 6 ÷ 12, even secondary aged pupils can be quite comfortable in identifying this as 6 divided into 12 with the resulting answer 2.

Changing role of teachers

With the complexities that the language and symbols of mathematics can involve, meanings need to be negotiated so that they fit with children's understanding but also with the conventions of more formal mathematics. There may be discrepancies between what pupils hear and what teachers hear, and between the meanings of what pupils say and what teachers say. The role of the teacher is to analyse to what extent meanings are shared and to identify how new/modified meanings are negotiated. In negotiating meanings, teachers will be mindful of the problems children will face if their understanding of numbers and operations is not extended to include multiple interpretations of the symbols, and flexibility in solution strategies.

In order to develop the skills needed to become successful problem solvers, children will need to

- realize that some words and symbols have multiple meanings
- become aware of the limitation of some meanings for words and symbols
- be convinced of the need to progress from more naïve interpretations to new terminology
- be aware that teachers' and other children's meanings may differ from their own
- be able to select appropriate interpretations of words and symbols and adapt procedures to match different types of problems.

Teachers will need to

- listen to children to assess existing meanings associated with words and symbols
- be aware that there may be discrepancies between teacher meanings and pupil meanings

- provide experiences that illustrate a diversity of meanings for words and symbols
- use language that has meanings for individual children as well as mathematical correctness.

Effective teaching involves strategies for interactions that are both responsive to the child's existing understanding and pro-active in negotiating new meanings. Opportunities to share their thinking with others will encourage children to reflect on the methods and language they themselves use and become aware of alternative interpretations and strategies.

Innovations in the classroom

With the changed emphasis from 'showing and telling' to engaging pupils in discussion and negotiation there have been classroom innovations that have had impact on the practices of teaching. Ways of thinking have become more exposed and there are innovations in the form of models and images that make it possible to share ideas and make calculation strategies easier to explain. Before progressing to look at some such innovations, consider again the calculation 6067 – 5970 illustrated above as one of the most complex subtraction calculations in 1937. This is a good example of a calculation that is formidable if the algorithm is used, and much easier using informal methods. A new innovation used in schools to support informal methods for calculating is the *empty number line* (a blank line with no markings) which can provide individuals with a visual image associated with their strategy. By visualizing the numbers 6067 and 5970 on an empty number line with the number 6000 between them, the calculation can be taken in two 'jumps': a jump of 30 from 5970 to 6000 and a further jump of 67. In this way the solution 97 is reached by a mental method, perhaps supported by an empty number line jotting.

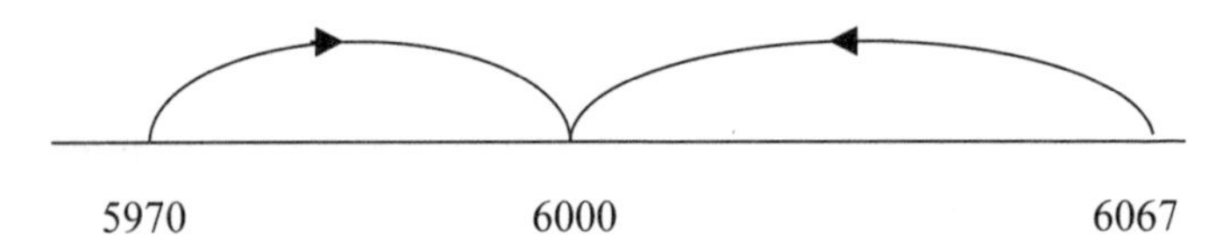

Figure 1.4 6067 – 5970 illustrated on an empty number line

There are other innovations in calculating like the *chunking* method used for division and some revived methods such as the *grid* method for multiplication that relate to particular ways of calculating. There are also diagrams such as the *percentage bar* and the *ratio table* that can be used to model a solution strategy providing a visual image to support thinking. These and other models and images will be considered in detail in later chapters.

New technologies

Perhaps the most substantial innovations in mathematics teaching are the technologies, from calculators to interactive whiteboards, that are now widely used in mathematics teaching. Not much more than a generation ago, all arithmetic in the workplace was undertaken by 'human calculators' who shared the same standard methods and recorded their working according to specific conventions. This was necessary as calculations were transferred from person to person and workplace to workplace. In society today, all important calculations are done by computers or calculators, from shopping bills to financial forecasting, and there are innovations like spreadsheets to support complex calculating. Society no longer has the same need for trained calculators, but needs individuals who can find different ways of approaching problems and can use their initiative in planning solutions. The National Numeracy Strategy reflects the need for changes in the curriculum when it introduces the word 'numeracy' as a 'proficiency which ... is more than an ability to do basic arithmetic ... it involves developing confidence and competence with numbers ... requires understanding of the number system, a repertoire of mathematical techniques, and an inclination and ability to solve ... problems in a range of contexts' (DfEE, 2001: 9). Teachers today have the responsibility for developing this confidence, competence and inclination for mathematics and using new technologies can support these objectives. Calculators, for example, can stimulate thinking and experimenting with numbers in a non-threatening way with immediate feedback on number patterns and calculations. Complex problems can be attempted with numbers in real contexts rather than the contrived numbers so often found in textbook exercises. Research shows that children will often try much harder examples when using calculators, and are particularly interested in exploring large numbers (Forrester, 2003). The way children use calculators appears to be different from the way they are used by adults, and individuals or pairs are presented with opportunities for learning through 'interrogation' as part of a game, perhaps, and through discussion or reflection on the output. Rousham (2004) identifies tasks with calculators designed to enhance thinking rather than simply used to do isolated calculations and talks of the 'feedback loop' he observed when children had extended access to calculators.

In addition to calculators, computers are available in most classrooms with spreadsheets as a standard feature for dealing efficiently with many number activities. Although spreadsheets were not originally designed for use in schools they are an excellent tool for exploring numbers and displaying results using a variety of diagrams including pie charts and bar charts. Perks and Prestage (2003) give a lot of advice about adapting the facilities to make them attractive and a useful tool for teachers to produce resources as well as a tool for children to use to explore number relations.

In addition to the standard features of a computer there is a growing wealth of software specially designed for classroom use, some from commercial publishers but also developed within local networks of teachers. This continues to be developed and internet searches will help teachers find particular resources for their own classroom. Websites for mathematics education are being created all the time and Way and Beardon (2003) outline effective ways to find mathematics education websites and to evaluate their suitability. Such resources can be used in conjunction with interactive whiteboards that are fixtures in many classrooms and provide opportunities for children to take a leading role in the development of ideas and actively participate in their learning.

Role of assessment

Finally in this chapter on the influences for twenty-first-century mathematics teaching, the role assessment has played must be considered. With the introduction of systematic testing, not only at the end of each Key Stage, but also with optional tests that have been widely used for preparation and practice, teachers have had to be mindful of the types of questions that have been used and the knowledge, skills and understanding they assess. At Key Stages 2 and 3 calculators have been allowed for one of the test papers and children have needed to identify where they can be helpful and how they are to be used. The Qualifications and Curriculum Authority have published very helpful reports analysing the way children have tackled problems and suggesting implications for teaching. These reports have included examples of effective solution strategies and support children's use of informal strategies rather than standard procedures that are poorly understood. Throughout this book use will be made of the examples and suggestions from these national reports.

In recent years the national tests have been typified by questions that will be answered more successfully through strategic thinking and making connections rather than disciplined application of an algorithm. Data has to be extracted from a word problem, pie chart or graph, and the result of a calculation has to be interpreted to fit the context of the question. No standard methods are required for calculating and many questions appear to invite informal approaches so that children can make their own sense of the questions. Assessment has always provided guidance to teachers about the strengths and weaknesses of their pupils, and with the focus on problem solving and non-standard questions they now have better access to their children's thinking than in the era when every child would use the same standard methods for calculating. Throughout the chapters of this book the focus will be on progressively developing children's number sense and examples will be used to exemplify the thinking that is expected of children today.

CHAPTER 2

Understanding Numbers

In learning mathematics one of the first things that children begin to build is their experience with individual numbers in their oral form. This will start as an eclectic set of isolated bits of information gathered from their experiences, such as ages, house numbers and various rhymes and stories. This knowledge will gradually build into a meaningful system that will continue to be extended when the need arises for new types of numbers. As they progress through school their understanding of numbers will not only be in contexts but in the relationships numbers have with each other and in the patterns that are created. Different characteristics will give names to certain numbers, for example *even*, *prime* or *square* numbers and the expectation is that children will have a wide vocabulary for numbers by the time they reach the end of primary school. In some number questions used in national tests no calculation is required but children's understanding of particular terms is assessed as seen in the sorting table from a Key Stage 2 national assessment question from 2004 (Figure 2.1).

Write a number **less than 100** in each space.

	even	**not** even
a square number		
not a square number		

Figure 2.1 A question from KS2 in 2004

From natural numbers to irrationals

When written symbols are introduced patterns become apparent for constructing numbers of any size with mathematical conventions for representing them. From the counting or *natural numbers*

1, 2, 3, 4, ... the introduction of zero and negative numbers extends the system to all whole numbers or *integers* . . . . , -3, -2, -1, 0, 1, 2, ... When parts of a whole are needed then fractions or *rational numbers* extend the system again. In the past vulgar fractions such as ½, ¼ and ⅛ were common, not least because our system for money and measuring found these in common usage. Since the introduction of decimalization in the last century it is more common to use decimal fractions such as 0.5 and 0.75. This has had impact on the school curriculum and the need to calculate with vulgar fractions is considerably reduced.

There are more numbers that will be explored in secondary school and in higher mathematics. The set of *real* numbers is made up from *rational* numbers (all decimals that can also be represented as fractions) and *irrational* numbers (decimals that are non-terminating and non-repeating) such as $\sqrt{2}$ or π which cannot be expressed as a ratio or fraction.

Understanding numbers through problem solving

As well as working with the more abstract patterns of numbers it is important at all stages of learning to work with numbers in many different contexts. In order to understand that mathematics is crucial for making sense of the world it has become more important to apply mathematics across the boundaries of many subjects and mathematical problem solving using the numbers that are appropriate for each different situation.

> Mathematics is of central importance to modern society. It provides the vital underpinning of the knowledge economy. It is essential in the physical sciences, technology, business, financial services and many areas of ICT. It is also of growing importance in biology, medicine and many of the social sciences. Mathematics forms the basis of most scientific and industrial research and development. Increasingly, many complex systems and structures in the modern world can only be understood using mathematics and much of the design and control of high-technology systems depends on mathematical inputs and outputs.
>
> (Smith, 2004: 11)

Through meaningful problems it is possible to identify the need for different numbers with their particular representations, and to provide contexts in which calculations have some purpose. This not only relates mathematics to experiences that may provide informal approaches to the solution of problems but also shows that mathematics is a practical subject that has many applications in the 'real world'. Meaningful contexts can include puzzles and games that may be constructed to incorporate the range of numbers that lie within the children's experiences, or may be everyday life problems that involve real numbers that extend and challenge their understanding. This will sometimes mean that the numbers involved are more complex than the examples found in books, but calculators can be used to support calculating and the sense that numbers have real purposes can be fundamental in motivating children.

Wherever you look there are numbers that provide useful information that can be of interest to children. From the earliest stages of nursery rhymes and stories, to the world of shopping catalogues and football league tables, there are numbers to be explored and investigated and given meaning. Smaller numbers are found in the number of pupils in a class, the cost of fruit and sweets, or the number of pages yet to be read in a book. Larger numbers are found on food labels, holiday guides and population data. Money and measures can be the basis for practical activities involving numbers in a useful context and will give children experiences that help link school to their home environment. 'Who can bring to school the heaviest potato?', or 'find out on the internet the population of their home town', or 'find the weight of an elephant' are all tasks that will give meaning to numbers. Working with planetary distances or sizes of molecules can motivate the introduction of different systems for representing numbers that go beyond the range of those in everyday use. Even in everyday use now are numbers of megabites or gigabites that some children understand better than their teachers. The advantage of working in real contexts extends beyond sense-making for the numbers, and meaningful questions arise introducing calculations that children may be interested in solving. These skills of using numbers to present and interrogate real data will provide the essential basis for understanding and succeeding in the technological society of today.

Words and symbols

Starting with the whole numbers, children come to understand the word and symbol patterns that govern the construction of all numbers. Through a process of counting they will come to memorize the number words and their particular order, and later learn the symbols associated with each number word. Counting is an essential skill that will underpin many calculating methods, and although counting to ten, or even twenty, may be done by pure memory, counting further will involve mathematical thinking as new numbers are constructed according to a strict pattern.

The representation system we use for numbers is complex and beautifully concise, and will take children many years to master completely. It begins with ten symbols: 1, 2, 3, 4, 5, 6, 7, 8, 9 and 0, called *digits* which, with the decimal point, will enable us to label all the counting numbers, no matter how big or small. The representations we use today are called Arabic numerals and they have been widely used across the world since the thirteenth century. This system was popularized by the Arabs but largely invented by Hindu mathematicians before the sixth century. The Hindu-Arabic system differs from its predecessors by introducing 'zero' as a 'place holder' and identifying symbolic representation for numbers by the way digits are ordered. The impact of the zero place holder is that different numbers can be represented using the same symbols, for example 202 and 220, which will be crucially different as recorded measurements. We refer to this system as a *place value* system and have rules for constructing a symbol for every whole number. Experiences with measuring distances or heights, and recording the results, will help to focus on the importance of placing the digits accurately. An understanding of the number system will be integral to calculation methods, whether the column calculations illustrated in the last chapter, or for patterns that can be visualized on a number line. The emphasis on splitting numbers into 'tens and units' does not reflect the more flexible approach that is needed for successful mental calculating and it is more effective to recognize that numbers can be split in many ways to suit different calculations (Beishuizen and Anghileri, 1998). It is more helpful, for example, to recognize that 26 = 25 + 1 when calculating 26 × 4 than to split 26 as 20 + 6, or that 98 is 70 + 28 when dividing by 7.

Occasionally today we continue to use the system of 'roman numerals' which is built with no place holder, so that more and more symbols need to be introduced as numbers get larger. Some roman numerals are quite compact, for example MMVI for 2006, but consider MCMXCIV as the representation for 1994. In later primary school, or early secondary school, when children are developing confidence with the Arabic number system, it can be both interesting and helpful to make contrasts with the rules for constructing numbers in other systems. There are many historic systems that may be studied such as those used by the Egyptians, Babylonians or Mayans, and an internet search will provide a wealth of information on these and many others.

Place value

In the early stages of writing numbers, children sometimes become confused between such numerals as 17 and 71, particularly as the name 'seventeen' rather invites the 7 to be written first. All of the number names between 'eleven' and 'nineteen' follow a different pattern to the one that is established, and becomes more stable when 'forty' is reached. Even the names for 'twenty' and 'thirty' differ subtly from the convention of adding 'ty' which is clear in six-ty, seven-ty and eigh-ty. This means that the early numbers have to be memorized by young children, and their representation as numerals matched to their spoken names. The 'oral and mental starter' to mathematics lessons provides the opportunity for counting practice and children can quickly master counting forwards and backwards, starting at different numbers, and 'skip' counting in twos, fives, tens and so on. As the symbols are recognized and number names learned, it is important to establish connections between the names, the symbols and a variety of visual representations of numbers. Even with small numbers, visualizations will exemplify a rich source of relationships that underpin many calculations. A collection of coloured beads on a string can be used to show that five and one more make six, or that six is double three, or three lots of two. An array of six counters on an overhead projector or an interactive whiteboard can be used to show the patterns; 5 + 1, 4 + 2, 3 + 3, 2 + 4 and 1 + 5. It is here that mathematical thinking begins as these patterns are observed and discussed, and although the full implications will not be appreciated initially, they are an important experience in building the connections that will make later number work meaningful.

There are many classroom resources, such as arrow cards and number lines, which are designed specifically to help children understand the construction and ordering of numbers. A bead string with different coloured beads in groups of ten has been suggested as an important visual representation for the pattern of numbers up to 100. With this resource it is particularly important to work around the decade boundaries (20, 30, 40 and so on), counting forwards and backwards, because the counting and the images together will form the basis for powerful mental methods (Beishuizen and Anghileri, 1998) and these will be explored in later chapters.

Other resources include a *hundred square*, starting either at zero or at one, with the number arranged in lines of ten. The advantage of starting at zero is that all the same decade numbers, the twenties, the thirties and so on, appear in the same line. Counting in ones is associated with movement along each line, while adding on ten, or subtracting ten, relates to a move down or up a column. This distinction can be confusing and more recently the empty number line has been introduced as an image for counting that is easier to visualize (Anghileri, 2006b).

The empty number line

In the following chapters, mental and written calculation methods will be introduced that involve visualization on an 'empty number line', that is a self-drawn line with no markings except those put on it by an individual. It may seem odd that this is considered to be an innovation in classroom teaching, but the important step of removing all marks from the number line allows this to be used flexibly by individuals as an aid to thinking beyond counting (Beishuizen, 1999). An empty number line can represent any range of numbers and illustrate the distances between them schematically rather than accurately. Returning to the example from the last chapter, 6067 – 5970, it can be seen that a number line starting at zero, or a number line with calibrations, may inhibit the thinking that can be associated with identifying a position for 6000 and then setting the other numbers in relation to this position.

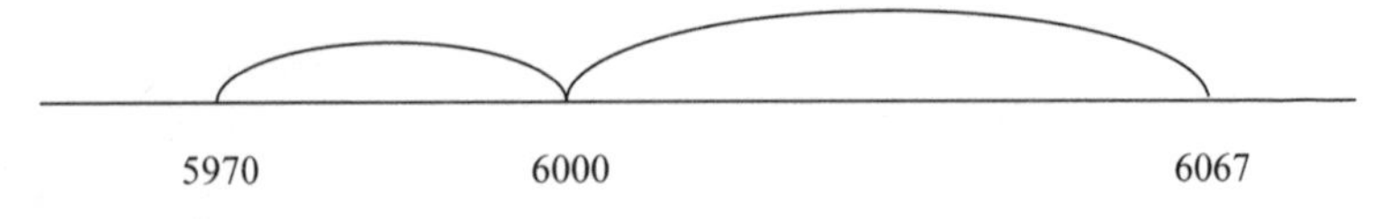

Figure 2.2 Locating 5970, 6000 and 6067 on an empty number line

This ability to identify relative positions of numbers is sometimes called *locating* numbers on a number line and is a skill that children can begin to build as soon as they know the ordering of numbers. Young children can be involved in a game of 'washing lines' where they each hold a number card and work together to find their correct positions in relation to each other. In this way, verbal counting and the ordering of symbols will begin to be associated in a way that can lead to the important skill of *visualizing* the number line. This can be a challenging activity for older children according to the number cards they are given, for example, ordering the numbers 0.1, 0.5, 0.15, 0.55 and 0.01 will need a lot of understanding of decimal numbers. A remarkable image is formed if the numbers to be located are the fractions $\frac{1}{2}$, $\frac{1}{3}$, $\frac{1}{4}$, $\frac{1}{5}$, $\frac{1}{6}$ and many children are surprised that these numbers are not evenly spaced, and that they occupy such a small part of the number line.

Ordering numbers up to 100 can be associated with the image of a bead string which, together with counting, can help children to understand the number positions on a number line. Positioning the decade numbers, 10, 20, 30 and so on, on an empty number line, will provide *benchmarks* around which other numbers can be placed. Discussing where numbers are positioned, for example that 29 is one before 30 and 28 is two before 30, will help to develop from counting skills into calculating methods for later addition and subtraction. In the later primary years, children will be familiar with the number names and positions for two-digit numbers and will be ready to extend to hundreds, thousands and bigger numbers. Positioning numbers around the benchmark 1000, or about 405000, or 1000000, will help to develop understanding of the structure of larger numbers.

Some recent assessment tasks at Key Stage 2 and at Key Stage 3 have focused on understanding of the number system with questions such as:

> What is the number half way between 30 and 80?

And

> Circle the number **nearest to 1000**
>
> 1060 1049 1100 960 899

Each of these questions can be tackled by visualizing the numbers on an empty number line. Another aspect of number is the idea of 'rounding to the nearest whole number'. This can be a very abstract activity for some children until it is related to a position on an empty number line. Then the equivalent question 'Which is the nearest whole number?' relates to a visual image and can make more sense. At a later stage this image can be used for rounding to one or two places of decimals.

Extending the number line

Numbers in context can provide a meaningful opportunity to extend the number line to include negative numbers, or numbers below (or to the left of) zero. There is no reason for the number line to extend horizontally rather than vertically although the convention for numbers extending horizontally from left to right fits with our conventions for reading and writing. While the non-zero whole counting numbers are called the natural numbers, when zero and the negative numbers are included the collection is called the integers. Conventional contexts for using negative numbers include temperatures, metres below sea level, or the more visual idea of a lift that goes below ground level. The convention for denoting a number that is 1 below zero is -1, the same as the operation denoting 'subtract 1' and this can cause confusion so that the alternative notation of $^{-}1$ is sometimes used. The number line is again a helpful image, with numbers to the left of zero being given a 'negative' sign while those on the right have a 'positive' sign.

Already, complexities arise with the notion that some numbers are 'smaller than nothing' and that $^{-}7$ is smaller than $^{-}5$, which is clear in the sense that $^{-}7$ is positioned further to the left of zero than $^{-}5$. Mathematical notation can be used to express these ideas using the symbols '$<$' meaning 'is less than' and '$>$' meaning 'is greater than'. So '$^{-}7 < {}^{-}5$' is read as '$^{-}7$ is less than $^{-}5$' and '$^{-}5 > {}^{-}7$' meaning '$^{-}5$ is greater than $^{-}7$'. At a later stage the symbols '$\leq$' and '$\geq$' will include the possibility that the two numbers are equal and will be used in algebraic expressions such as $x \leq 5$. Mixed expressions such as $0 < n \leq 5$ will include the possibilities that n is any of the whole numbers 1, 2, 3, 4 or 5 (or later, any of the numbers between 0 and 5). Once negative numbers can be positioned on the number line, addition and subtraction can be identified with jumps to the right and to the left respectively but jumping a negative number of steps reverses the operation. In this way, adding $^{-}5$ will be equivalent to subtracting 5. Subtracting $^{-}5$ will be the same as adding 5.

Building more numbers

In dealing with decimal representation, numbers have been introduced that lie on the number line between the whole numbers. This has increased the collection of numbers from the integers (all positive and negative whole numbers and zero) to the set of real numbers (all decimal numbers). When introducing decimal numbers it is worth noting that children will already have extensive experiences through representations used for money and measures. This can present a considerable handicap if the misconceptions associated with this representation are not addressed. Symbols like £0.99 and £2.05 are likely to be interpreted as 'ninety-nine pence' and 'two pounds and five pence' without any understanding that the first decimal place represents tenths and the second represents hundredths. Research has shown that secondary age children can still think that 3.4 and 3.04 represent the same number (see Nickson, 2004 for references). It will be some time before the systems for money and the positions of decimals on a number line can be reconciled. When introducing the decimal system it is helpful for children to have a purpose in using the notation for tenths and hundredths. Measuring distances in metres can provide one such opportunity as more accurate measurement can be achieved by introducing tenths. The decimetres need not be named initially as decimetre sticks are used alongside metre sticks to measure and record distances with the decimal numbers, for example 3.4, to indicate three whole metres and four tenths. After considerable experience with tenths, a hundredth unit, in this case a centimetre, can be introduced for even more accuracy and the relationships between tenths and hundredths explored.

Other contexts to give meaning to tenths are those involving capacities of containers measured in litres and tenths of a litre. Using special containers for a litre and a tenth of a litre (decilitre) the children themselves can calibrate a variety of containers, marking calibrated scales and using them for measuring. Recording and comparing the amount held in different drink cans and bottles, or shampoo containers can be linked with ideas about economic buys and value for money that are often fascinating and real questions for children.

Weighing quantities for cooking or preparing animal feeds, using kilograms and grams, will provide further experiences for using hundredths, although the unit for a tenth of a kilogram is not often used. Using different calibrated scales, including linear, circular and digital displays, and interpreting the readings, will provide necessary experiences for practical applications and for answering those assessment questions that frequently appear on national tests, and examples will be found in the chapter on measuring.

Some special numbers

As children become aware that patterns are very important for understanding numbers, both in a verbal sense, and a visual sense when looking at collections of objects, they can begin to recognize that in some collections the objects can be paired while in others there will always be an 'odd' one left. The numbers that give rise to a paired collection, 2, 4, 6, 8 ... are called the *even* numbers that may also be recognized as the multiples of 2. All the other counting numbers, 1, 3, 5, 7, 9 and so on, are called *odd* numbers. A lot of mathematical thinking can be explored in determining whether a number such as 27 will be even or odd. As well as rearranging sets of objects, a simple (non scientific) calculator can be used to generate even numbers by pressing the buttons '2', '+',

'+' and then '=' repeatedly. Children can be engaged in predicting whether 27 will appear, or 99, or 150. If they are able to recognize and describe the pattern that all even numbers end in 0, 2, 4, 6 or 8 then they have made an important discovery. This type of activity that encourages children to observe patterns and to describe what is happening is central to developing their mathematical thinking. Engaging in such activities with focused discussion will be a more powerful learning environment than one in which the teacher is expected to pass all knowledge on.

Some more special numbers are the *square* numbers 1, 4, 9, 16, 25 and so on, whose collections can be arranged as squares. In upper primary school these will be recognized as 'numbers multiplied by themselves', for example $2 \times 2, 3 \times 3, 4 \times 4, 5 \times 5$ and new notation as 2^2, 3^2, 4^2, 5^2 can be introduced. The use of the smaller number, or *index*, is another important development that will be used extensively in secondary school. Confusion can arise for some individuals if the first index to be introduced is $2^2 = 2 \times 2$. Better to start with 3^2 or 5^2 or even 10^2 as a different way of writing 100. *Triangular* numbers can be arranged as triangles

Another interesting collection of numbers is the *prime* numbers: all the numbers that do not appear in any multiplication tables other than as multiples of 1. (All other numbers that are in multiplication tables are called *composite* numbers.) These prime numbers have fascinated mathematicians throughout history as they do not have any recognizable pattern when ordered on a line or in any other arrangement. The mathematician Eratosthenes devised a method, called the *Sieve of Eratosthenes*, that children can reproduce for finding primes. On a hundred square, first cross through all multiples of 2 and then all multiples of 3, 4, 5 and so on until all composite numbers are crossed through. The numbers that are left are the prime numbers and it can be seen that, even up to 100, they appear at different intervals throughout the table in no particular pattern.

Extending notation for large and small numbers

The notation for large numbers no longer has the single convention of a comma separating the thousands and the millions although these are helpful. If you take a look around at all the uses for large numbers that exist in society today, from phone numbers and passport numbers, to the numbers found on all types of labels (tin cans to chocolate bars), they adhere to conventions that are meaningful in a particular context. Since these are the numbers that children encounter in everyday life, there is a need to review the different forms that numbers can take and make decisions about the most useful interpretations. Involving the children in decisions about the uses and interpretations of numbers develops important life skills and an understanding that number sense is a practical necessity. Projects such as comparing the nutritional values of different breakfast cereals, or the populations in different countries, or distances in global travel, will all involve meaningful use of numbers in real contexts.

There is a scientific convention for particularly large and small numbers, called *standard form*, which some children may like to explore in the later primary and early secondary years. This involves powers of ten with 10^2 representing 10×10 or 100, 10^3 representing $10 \times 10 \times 10$ or 1000 and so on. In a similar way, 10^{-1} represents $\frac{1}{10}$ or 0.1, 10^{-2} represents $\frac{1}{10 \times 10}$ or 0.01 and 10^{-8} would represent 0.00000001. It is not so useful to write 650 as 6.5×10^2 but the expressions 6.5×10^8 for 650000000 or 6.5×10^{-8} for 0.000000065 make concise representations that are more easily interpreted than numbers with very many zeros. The usefulness becomes clear when distances around the globe, or the size of atoms are involved. There are even some fascinating number names for very large numbers that have become more significant in today's technological age, for example a *billion* for a million million or 10^{12} (called a *trillion* in the USA) and a *googol* for the number 10^{100} (1 with 100 zeros).

Much of the discussion so far in this chapter has involved number properties that can be very abstract and pointless for some children. When numbers are used in context there is better motivation to understand their meaning and use them appropriately. Many examples of numerical data are accessible through the world wide web and children can be involved in finding their own numbers to work with. This presents opportunities for differentiation as the less confident children can be helped to find meaningful tasks with numbers they are confident with, while more confident children can be supported to extend their knowledge of numbers.

Numbers in society

Before leaving the representation system for numbers it is worth discussing further the many different conventions in operation in society today. The use of a comma, for example, to separate the thousands in 3,000 is no longer universal and a comma is common in continental Europe for representing decimal numbers like 3,5 (for 3.5 or 3 wholes and 5 tenths). A mobile phone number or a score in a computer game will use their own conventions and children can be very adept at using and understanding the purposes of such different systems just as they have become competent with using thumbs and text messaging. What is important is that children are exposed to many varied uses for numbers and come to recognize the power they can gain in mastering and using them.

CHAPTER 3

Making Sense of the Operations

In the last chapter it was noted that one of the first ways children develop an understanding of number is through subdividing sets of objects in different arrangements. A collection of six objects, for example, can be arranged as 5 and 1, 4 and 2, or as 2 sets of 3. The visual and tactile experiences gained will form the foundation for ways of manipulating numbers and are crucial for understanding the place-value system that we use for recording. Before the operations are introduced formally with the written symbol, however, young children will need extensive experience of re-organizing collections and talking about the way numbers are made. Words such as 'more than', 'less than', 'and' and 'take away' are part of everyday vocabulary which will play an important role when addition and subtraction are formalized. Ideas of equal sets repeated several 'times', 'sharing' into 'equal groups', sometimes with some 'left over', will provide imagery and words that are important for multiplication and division. This ability to partition numbers in a variety of ways will provide the key for efficient ways of calculating, for example, $98 \div 7$ can be effectively calculated if 98 is seen as $70 + 28$ rather than $90 + 8$.

As well as using collections of objects, more abstract patterns such as rhythms in music or movement can provide opportunities for counting. The rhythmic counting '1, 2, **3**, 4, 5, **6**, 7, 8, **9** . . . ', with emphasis on every third counting number, will be good preparation for the multiplication pattern of 3s. Counting forward and counting back will provide strategies for addition and subtraction that will later be refined and made more efficient. Practical activities and oral work will be essential at all stages as the imagery and verbalizing can be helpful in supporting the more abstract symbolizing that will follow. Government guidance for teachers proposes that:

> Early practical, oral and mental work must lay the foundations by providing children with a good understanding of how the four operations build on efficient counting strategies and a secure knowledge of place value and number facts.
>
> (DfES, 2006a: 2)

Establishing connections

Using a variety of language to talk about the activities children are engaged in will prepare the foundation for the connections that underpin all the arithmetic operations and then help develop a 'feel' for numbers. In school the 'four operations' are often learned as distinct ways of representing different calculating methods, but in higher mathematics there is only a distinction between additive operations and multiplicative operations, with subtraction and division included as 'inverse' operations. This is because the same three numbers are involved in both addition and subtraction, for example, 10, 6 and 4 are related in the expressions:

$$10 = 6 + 4 \qquad 10 = 4 + 6 \qquad 10 - 6 = 4 \qquad 10 - 4 = 6$$

All of these come from the understanding that 10 is made up from 6 and 4 and each symbolic representation gives a different way of expressing this relationship. For developing additive thinking the focus will need to be on such relationships rather than procedures for calculating. This is a departure from past traditions where the algorithms for calculating have been paramount.

Similarly, number triples are related by multiplication and division, for example, the numbers 12, 3 and 4 are all related in the expressions:

$$12 = 3 \times 4 \qquad 12 = 4 \times 3 \qquad 12 \div 4 = 3 \qquad 12 \div 3 = 4$$

When these expressions are put into words they can trigger very different images and procedures, for example '3 lots of 4' will look very different from '4 times 3', and it is the teacher's role to help children to not only solve each calculation, but to understand the connections. This understanding can be enhanced when the idea is introduced that 3 and 4 are *factors* of 12, giving children the language for discussing different ways in which multiplicative relationships can be expressed. This will be discussed more fully in Chapter 5. Some children come to understand these relationships early in their mathematical experiences and computation becomes flexible and easier than the calculations undertaken by those who see the operations as distinct procedures. Research has shown that the most effective teachers focus on these relationships rather than emphasizing practice of independent procedures. In her results Heirdsfield (2005) found that the two overarching issues for supporting learning are *establishing connections* and *encouraging strategic thinking.*

Developing additive thinking

As addition and subtraction provide different ways of expressing a relationship among three numbers, it becomes increasingly important to identify appropriate processes and vocabulary in order to understand these operations. Starting with numbers in contexts has the advantage that everyday language will be a helpful guide to processes for finding solutions. Symbolic representation will then provide succinct shorthand for expressing the relationship as pupils are encouraged to make a record of their problem solving. Rather than using the symbols to introduce the operations it is better for children to use them to record their thinking and their findings. In this way they will see that the same symbolic expression can be used to represent problems with different structures and that the symbols require many different interpretations.

Some of the complexities that arise have been analysed in research, for example the different problem structures that can embody addition and subtraction including those that involve actions and those that are static situations (Table 3.1).

Some of these problems will be easier to solve as the actions for a solution are embedded within the context, for example the problems that involve giving. In other cases difficulties arise because the situation is static or because starting information is not given. Children will need to experience situations and problems in order to develop an understanding of addition and subtraction that encompass all these variations. They will need to develop familiarity with the language structure involved, in particular 'more than' and 'less than' (or 'fewer than').

Table 3.1 Types of addition and subtraction word problems (Riley et al., 1983)

Types of word problems (from Riley, 1981)	
Action	Static
CHANGE Result unknown 1. Joe has 3 marbles. Then Tom gave him 5 more marbles. How many marbles does Joe have now? 2. Joe has 8 marbles Then he gave 5 marbles to Tom. How many marbles does Joe have now? Change unknown 3. Joe has 3 marbles. Then Tom gave him some more marbles. Now Joe has 8 marbles. How many marbles did Tom give him? 4. Joe has 8 marbles. Then he gave some marbles to Tom. Now Joe has 3 marbles. How many marbles did he give to Tom? Start unknown 5. Joe has some marbles. Then Tom gave him 5 more marbles. Now Joe has 8 marbles. How many marbles did Joe have in the beginning? 6. Joe has some marbles. Then he gave 5 marbles to Tom. Now Joe has 3 marbles. How many marbles did Joe have in the beginning. EQUALIZING 1. Joe has 3 marbles. Tom has 8 marbles. What could Joe do to have as many marbles as Tom? (How many marbles does Joe need to have as many as Tom?) 2. Joe has 8 marbles. Tom has 3 marbles. What could Joe do to have as many marbles as Tom?	COMBINE Combine value unknown 1. Joe has 3 marbles. Tom has 5 marbles. How many marbles do they have altogether? Subset unknown 2. Joe and Tom have 8 marbles altogether. Joe has 3 marbles. How many marbles does Tom have? COMPARE Difference unknown 1. Joe has 8 marbles. Tom has 5 marbles. How many marbles does Joe have more than Tom? 2. Joe has 8 marbles. Tom has 5 marbles. How many marbles does Tom have less than Joe? Compare quantity unknown 3. Joe has 3 marbles. Tom has 5 more marbles than Joe. How many marbles does Tom have? 4. Joe has 8 marbles. Tom has 5 marbles less than Joe. How many marbles does Tom have? Referent unknown 5. Joe has 8 marbles. He has 5 more marbles than Tom. How many marbles does Tom have? 6. Joe has 3 marbles. He has 5 marbles less than Tom. How many marbles does Tom have?

Working with different symbolic representations

As children work from contexts to their representation symbolically, they will begin to see patterns emerging with the same numbers cropping up in many different relationships. When they understand the way the symbols '+' and '−' can be used to record activities and problems involving addition and subtraction, alternative representations for the same number relations can be introduced so that connections are emphasized. Initially this will involve benchmark facts such as the *number bonds of ten*, that is, the pairs of numbers that are added to make ten. Later, more complex calculations can be represented in a variety of ways. With the calculation $25 + 46 = 71$, for example, there are associated many possible problems as well as the standard $25 + 46 = \square$:

$46 + \square = 71$	$\square + 46 = 71$
$25 + \square = 71$	$\square + 25 = 71$
$71 - \square = 46$	$71 - \square = 25$
$\square - 25 = 46$	$\square - 46 = 25$

It will be helpful if the reader now takes a few minutes to read aloud each of the above representations and identifies which would be the most complex to solve. The first two may look very similar but children find $\square + 46 = 71$ much more difficult to solve than $46 + \square = 71$. Children feel more comfortable when they can start with 46 and add more on until 71 is reached. The expression $\square + 46 = 71$ is referred to as 'start unknown' and many children cannot think of a strategy for finding the number onto which 46 must be added. Where this type of question is embedded in a realistic situation the context can sometimes help in finding a solution. For example, 'Jon started with some money and was given 25p more. Now he has 71p. How much did he start with?' This may still appear daunting but teachers can ask leading questions like 'Did Jon have more than 20p?' and 'Did Jon have as much as 50p?' to model a possible strategy. This strategy, known as *trial and improvement*, is not the most efficient but may provide an effective approach for many children. It is important that the teacher does not always provide the context as children need to learn to identify abstract calculations with some real problem or imagery for themselves.

It can be seen from this analysis that a good understanding of addition and subtraction comes from extended explorations of different problem structures rather than a set of abstract procedures. Working together with the teacher, discussing different solution strategies and recording results, children will gain experience of the number triples that are involved and patterns will begin to emerge. These patterns will provide the number facts and connections that are essential for calculating, and classroom activities can focus on the visual images that support memorization.

Images that support additive thinking

Visual patterns provide the key to many mathematical relationships and the spatial arrangements of objects will be the first patterns encountered. Children first become aware of the numbers that are *doubles*, for example 3 + 3 = 6, and these can be extended to *near doubles*, 3 + 4 = 7, so that already connections are being made. Coloured cubes can be made into pattern sticks or a bead frame with coloured beads can be used to show relationships among numbers up to ten and these can be reinforced by colouring strips of ten squares or bead patterns on paper. At this stage a very important ***law of arithmetic***, the ***commutative rule*** for addition, can be seen: when adding two numbers the same total is reached if the numbers are reversed. This is not obvious to all children and will need to be the focus for systematic discussion as it will provide an efficient shortcut for many calculations. In the Netherlands the number bonds of ten are described as 'hearts in love' so that '8 and 2' and '2 and 8' assume equal importance in memorizing the number fact. (This emotive description seems to have more appeal than the rather stark 'number bonds'.)

In Chapter 2 a number line was introduced to represent the counting numbers with the idea of addition as a jump from one number to another. This can be used to explore the commutative rule although the images for 8 + 2 and for 2 + 8 will look quite different. The number line is associated with counting and it will be seen how much more efficient it will be to count on 2 more from 8 than 8 more from 2, and *starting with the larger number* will be identified as a step towards better efficiency. When the calibrations are removed to give an 'empty' number line, unitary counting is discouraged to encourage the facts to be memorized. By jumping forwards and backwards on the empty number line and verbalizing the results the relationship between addition and subtraction becomes evident. At the same time that they learn that 8 and 2 make 10, and that 2 and 8 make 10, children can learn that 8 is 2 less than 10 and 2 is 8 less than 10. They will also have an image to associate with these connections.

It is important to note that the commutative rule does not hold for subtraction. Whereas two numbers to be added can be reversed to give the same total, when the operation is subtraction the rule no longer applies. A well-established misconception concerns the application of the commutative rule to subtraction, particularly in traditional column arithmetic as reports on national examinations illustrate (Figure 3.1).

Question A5 assessed children's ability to subtract 198 from 309. Just over half of children working at level 3 answered correctly. The most common answers given by a fifth of children working at level 3 were 291 or 201. An answer of 291 was possibly reached by using an inappropriate strategy of subtracting the smallest digit from the largest digit throughout. An answer of 201 was possibly reached because children thought that the larger digit could not be subtracted from zero.

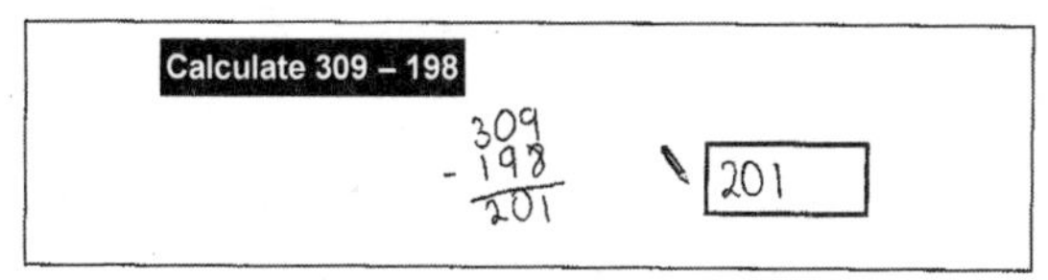

Test A, Q5, example of an incorrect response.

Some children successfully used informal methods of counting on to the next multiple of 100. The use of number lines is an effective informal method and highlights the notion of 'difference', as this example of a self-drawn number line demonstrates.

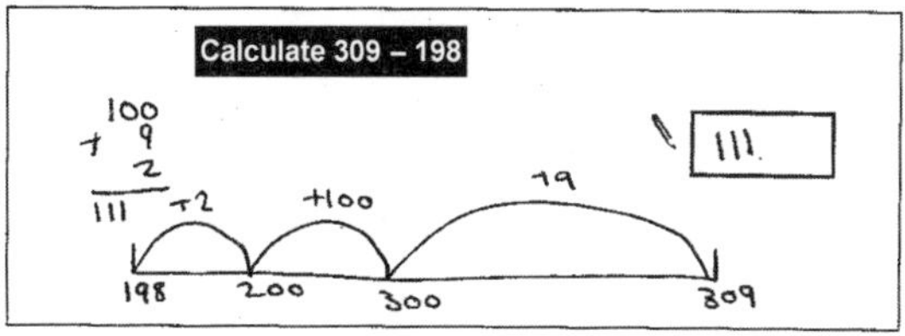

Test A, Q5, example of a correct response.

Figure 3.1 National test report on children's methods for calculating 309 – 198

When more than two numbers are to be added there is another useful rule that can make calculating more efficient and that is clearly illustrated with practical materials. If three numbers are to be added, for example 8 + 7 + 3 then a second ***law of arithmetic***, the ***associative rule*** for addition, enables the latter two numbers to be added and then this total added to the first: 7 + 3 = 10 and then 8 + 10 = 18. (Note that the order in which the numbers are added does not change but the pairs to be 'associated' first can be selected.) Extending this rule to a whole list of numbers to be added and using it in conjunction with the commutative rule means that pairs can be selected and added using known facts rather than the numbers added in their particular order, for example,

17 + 25 + 33 + 15

may be seen as 50 + 40 by pairing 17 + 33 and 25 + 15. This was a very efficient shortcut in past times for human calculators who became very effective at identifying quick ways to do long calculations.

Once again, the application of the associative rule does not follow for subtraction. An expression such as 20 – 7 – 3 will give a wrong total if 3 is first subtracted from 7. In such expressions the use of brackets removes any ambiguity as 20 – (7 – 3) gives a different result from (20 – 7) – 3. Removing the brackets altogether requires a convention to be followed where subtraction takes place strictly from left to right unlike the rule for addition. This gives a very good purpose for introducing brackets as invaluable symbols in mathematics that will remove ambiguity and reduce the number of conventions to be memorized.

Using a calculator to explore addition and subtraction

As soon as the symbols for addition and subtraction are introduced, a simple four-function calculator can be a helpful aid to thinking. The government guidance proposes that 'the calculator can be an effective teaching and learning resource . . . [to be] used with children in all age groups across the Foundation Stage, Key Stage 1 and Key Stage 2' (DfES, 2006d: 1). At Key Stage 3 children are expected to know how to use the laws of arithmetic to support efficient and accurate mental and written calculations, and calculations with a calculator and to recognise that the answer on a calculator may be inexact. When a scientific calculator is introduced the bracket function can be used for calculations such as

a. $7.6 - (3.05 - 1.7)$ b. $\frac{8.4 - 3.7}{8.4 + 3.7}$

(DfEE, 2001: 108)

Using a calculator, patterns of numbers can be generated, for example starting with any number and repeatedly adding a constant. Pressing '3', '+', '+' and then the '=' button repeatedly will generate the pattern of multiples of 3 which children will often be happy to extend well beyond 30 or even 100. Talking about this pattern, for example why 30 and 60 and 90 are included but not 20 or 40 or 50, can generate thoughtful discussion. They may also observe that 100 is not featured and that the numbers jump from 99 to 102. This is part of the 'feedback loop' referred to in Chapter 1 and typifies observations that will add to an individual's growing awareness of numbers and help make connections that will be important in later calculating. A calculator is one of the best ways of exploring the pattern that emerges when 10 is added to, or subtracted from, any number. Playing in pairs, one child could enter any three-digit number and see if the other can tell what the result of adding or subtracting 10 will be. In this game numbers such as 201 or 399 can be quite challenging. For pupils in Key Stage 2 decimal addition and subtraction can involve repeatedly adding or subtracting 0.1 or 0.25 from any starting number. It is helpful to support such an activity with jumps illustrated on a number line so that the pattern of numbers is related to a visual image that can help understanding.

Another important use of the calculator involves entering and interpreting data relating to money and measurements. Adding quantities such as £3.75 + £2.15 will give an output of 5.9 which requires understanding in interpreting, while subtracting 375 metres from 2 kilometres will need thought about what to input. These ideas will be discussed more fully in Chapter 6. All the skills developed with a calculator do not reduce the necessity for children to be good at calculating mentally or with pencil and paper. Making mathematics meaningful and developing skills necessary for thriving in tomorrow's technological society means that every individual needs to be competent and confident using a calculator as well as efficient with mental and written calculating. They need to know *when* to use a calculator and *how* to use the different facilities for complex calculations.

Progressing to multiplicative thinking

When children are aware of some of the benchmark facts for addition and subtraction, their number sense can be extended to multiplication and division. Initially they will become aware of *doubles* as a result of *2 lots of* the same number and the related *halves.* The numbers constructed in this way, double 1, double 2, double 3 . . . , can be identified with *counting in 2s,* and will later be identified as *multiples of 2.* Counting in 3s can be related to the images of equal sets of 3 and similarly equal sets of 4, 5 and 10. To develop from additive thinking individuals will need to be able to count how many sets there are as well as how many objects in each set, and here the word *times* may be introduced to replace *lots of.* It is the language with 'lots of', 'equal sets', 'times' and 'how many times?' as well as the ability to count equal sets that extends thinking from addition to multiplication and division. As with addition and subtraction, multiplication and division exist as inverse operations that relate to the same number triple, and multiplicative thinking goes beyond identification of the independent operations.

In order to understand multiplication and division children need to make connections between verbal, visual and symbolic representations, and come to appreciate that multiplication is more than just repeated addition and that division is more than just sharing. Repeated addition makes little sense for explaining the meaning of 0.3×0.4 and sharing does not easily relate to the division $6 \div \frac{3}{4}$. In order to understand these expressions alternative meanings need to be given to the operations and appropriate visual imagery introduced. In Key Stage 3 both multiplication and division are related through the idea of a *ratio* and this is embedded in *proportional reasoning* which will be the basis for particular types of problem solving. This will be considered in Chapter 5 on multiplicative thinking and in Chapter 7 on fractions, decimals and percentages.

Counting patterns

Ideas of multiplication and division can be introduced initially through counting in 2s, 3s, 4s and so on, sometimes called 'skip counting', enjoying and memorizing each pattern of numbers as a verbal string. This can be reinforced by patterns generated on a calculator. DfES guidance advocates that

> Using the constant facility on a calculator to add 2s, 5s and 10s repeatedly to 0 is a way to derive the multiples and can help children identify the pattern of the numbers in the sequence.
>
> (DfES, 2006b: 2)

This enables children to build familiarity with certain number patterns but gives them little understanding of how they relate to the formal operations. In the past, chanting of such number patterns has been dismissed as playground activity, and not associated with a classroom activity. In particular this has not been related to the 'times tables' that have been so important. But these patterns *are* the times tables if the children learn to use them effectively, and can build multiplication facts far beyond the conventional tables. The rhythm and the word patterns provide many clues to the structure of the multiples, for example the very obvious pattern for the multiples of five: '*five, ten, fifteen, twenty, twenty-five, thirty, thirty-five* ... ' which is explained by the fact that five is half of ten. This is an example of a pattern that can be continued indefinitely to give multiples well beyond tables facts. Keeping track on fingers while skip counting will give ready access (although rather inefficient) to multiples up to ten times. The pattern for multiples of two is also very appealing and quickly learned by most children. Every other number in the pattern of twos gives the multiples of four, providing a connection that will be useful for multiplication and division calculations. By illustrating jumps of two and jumps of four on the same line this relationship can be illustrated visually (Figure 3.2).

Where the pattern is less obvious, as in multiples of three, or almost lacking altogether, as in the multiples of seven, it makes the facts less easy to remember, and multiples of seven are often the last to be learned.

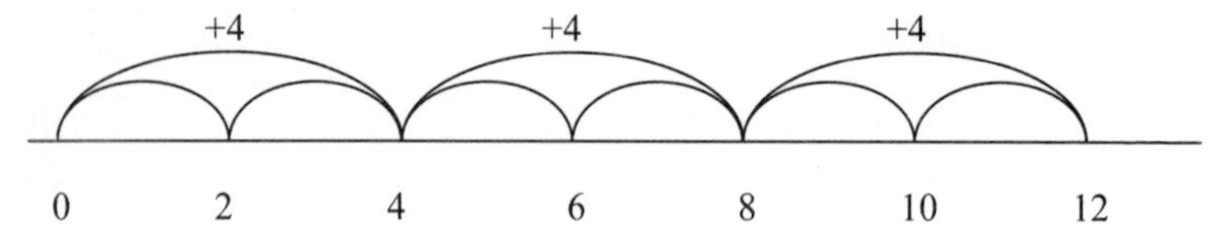

Figure 3.2 An empty number line showing jumps of two and jumps of four

In the past, learning of multiplication 'tables' meant memorising a particular chant, whether '*one three is three, two threes are six, three threes are nine, four threes are twelve . . .* ' or '*three ones are three, three twos are six, three threes are nine, three fours are twelve . . .* '. This makes the number facts accessible to some children but there are many who find such memorization impossible. There are also many who are able to recite these tables but who cannot use them meaningfully. It is true that children who do not have ready recall of multiplication facts will be disadvantaged in later calculating, but there are different ways of learning multiplication meaningfully. Where facts are not quickly recalled, as is often the case with multiples such as 7×8, individuals need the confidence to derive new facts from those they know, such as $7 \times 7 = 49$, so that 7 more can be added to give $7 \times 8 = 56$. In addition to the number patterns it is most important to establish the connections that will reduce memory load and make the derivation of many new facts possible. Doubling will generate new facts and highlight links between tables such as the twos and fours, threes and sixes, fours and eights and fives and tens. It will also give access to multiplication by numbers such as 16 or 20. Such links are central to understanding multiplication and will provide effective strategies for later mental calculating.

Images for multiplication and division

As well as verbal patterns for memorizing the facts, a range of different images will help children understand multiplication and its relationship to division. Repeated sets that are identified with equal jumps on a number line will initially provide an image that matches verbal skip counting. For investigating numbers that are not included in a particular counting pattern the image of *close numbers* and consequent ideas of *remainders* can be introduced and this will help children develop a strategy for division. The number 21, for example, will not appear in the *pattern of fours* but is one unit away from 20 giving rise to the fact that '*21 is 5 fours and one left over*'. The number 19 is also *close* to 20 but the remainder on division by four must be found by going to the multiple *before* 19, so that '*19 is 4 fours and three left over*'. This often causes confusion and the number line image with jumps illustrated will provide helpful imagery to support discussion. Recording these ideas can involve partitioning numbers into meaningful chunks for calculating so that when dividing by four, 21 = 20 + 1 while 19 = 16 + 3.

The array model

More difficult to understand from jumps on the number line is the commutative rule for multiplication which allows the pair of numbers in a multiplication (the *factors*) to be reversed to give the same total (the *product*). To give equal weighting to each of the factors an illustration of objects arranged in an *array* will be a helpful image. This will later be developed into a rectangle on squared paper giving meaning to the *area* aspect of multiplication. In describing the number relationships involved in an array the same visual image will need to be considered in two distinct ways so that a pattern can be seen as '5 lots of 3' and at the same time '3 lots of 5'.

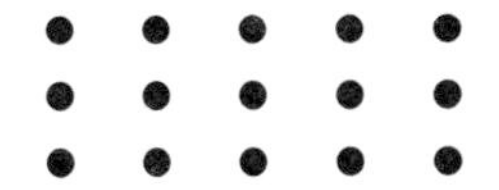

This will be obvious to adults but may take some time for children to master who may be helped by turning the page to see the image in a different orientation. When symbolic recording is introduced this array can be symbolized as 3 × 5 or as 5 × 3 which illustrates the commutative rule for multiplication, $a \times b = b \times a$.

Once the commutative rule is understood it will reduce by half the multiplication facts to be learned and will provide more efficient ways of calculating. For example 20 × 2, or 20 lots of 2 can be calculated as 2 lots of 20. It is important to make explicit to children that the commutative rule is not satisfied for division and

2 ÷ 20 cannot be replaced by 20 ÷ 2 although this is a common misconception. In a research project with 11 and 12 year olds the majority of children thought the answer to 6 ÷ 12 was 2 and read the calculation as '6 divided into 12' (Anghileri, 1995). Some time will be needed to discuss the meaning of the symbols and the rules that can be used.

When formal expressions for multiplication are introduced there will be ambiguity in interpreting them as long as children rely on imagery that does not involve the commutative property. The interpretations '20 times 2' and '20 multiplied by 2' will relate to '2 + 2 + 2 . . . + 2' twenty times and '20 + 20' respectively. Both interpretations are commonly used and children must develop flexibility in using the words and procedures that make most sense of the calculation. The array can be adapted to an array of squares or a *grid* and this is sometimes referred to as the *area* model for multiplication which will be discussed in Chapter 5.

When a fraction or decimal is involved, as in '½ × 15' or '12 × 0.2', it may be easier to use the whole number as the multiplier regardless of the interpretation given to the symbol. In each of these cases the area model with '15 lots of half a square' or '12 lots of 0.2' illustrated on a square give meaning to the calculations. This can later be related to jumps of ½ in the first case or 0.2 in the second case on an empty number line (Figure 3.3).

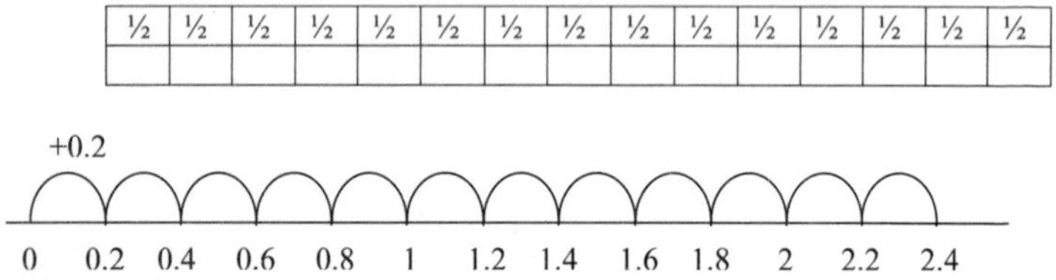

½	½	½	½	½	½	½	½	½	½	½	½	½	½	½

+0.2

0 0.2 0.4 0.6 0.8 1 1.2 1.4 1.6 1.8 2 2.2 2.4

Figure 3.3 15 lots of half a square and 12 lots of 0.2

Further meanings will be needed for developing work with fractions and decimals and these will be included in Chapter 7.

Another important law of arithmetic for multiplication is the associative rule. As with addition, this enables three numbers to be multiplied by choosing the appropriate pair to multiply first and this is shown clearly when brackets are used, for example (7 × 5) × 2 = 7 × (5 × 2). While some children will proceed in order multiplying 7×5 first to get 35 and then doubling it, others may 'spot' that it is easier to do the calculation as 7×10. Although this rule is not always made explicit to children it is used extensively, for example the calculation 6×40 can be related to the fact that $6 \times 4 = 24$ when it is considered as $6 \times (4 \times 10) = (6 \times 4) \times 10$.

Factors and multiples

Further experiences for developing multiplicative understanding involve chunking numbers in a variety of ways to show the multiplicative relationships involved.

The number 36, for example, can be 'chunked' (*factorized*) in the following ways:

$4 \times 9 \quad 2 \times 18 \quad 3 \times 12 \quad 2 \times 2 \times 9$

$3 \times 3 \times 4 \quad 3 \times 3 \times 2 \times 2 \quad 6 \times 6$

All of the numbers generated in this way, 2, 3, 4, 6, 9, 12 and 18, are the *factors* of 36 (together with 1 and 36) and where the numbers can be split no further, as in $2 \times 2 \times 3 \times 3$, this is described as the *prime factorization* of 36. A good activity for finding factors is to use squared paper and make all the rectangles with a given area, in this case 36. Along with a rectangle that is 1×36 all the rectangles will match the factorizations into a pair of numbers that are possible. Another activity that can be extended for finding the prime factorization involves making a tree diagram (Figure 3.4).

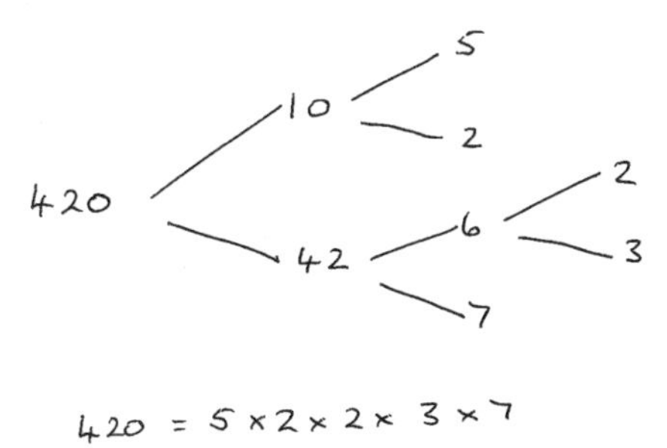

Figure 3.4 A tree diagram showing prime factorization of 420

A very *close* number, 37, is a *prime number* and has only the factorization 1×37, with 37 and 1 its only factors. Through continued experiences with multiples and factors children will develop the number sense that enables them to 'spot' numbers that have many factors and numbers that are likely to be prime. Finding factors of even very large numbers can be made easy with the use of a *spreadsheet* which will do many calculations very quickly. Using the formula facility to divide a column of numbers by 2, then 3, then 5 . . . at the stroke of a button the factors which divide any given number exactly can be found.

Also useful are *divisibility tests* that give some characteristics of a number that will tell immediately what some of the factors are. The first, and most obvious, divisibility tests are those for 2 and 5. Every even number is divisible by 2 and these are the numbers that have a final digit that is 0, 2, 4, 6 or 8. All *multiples* of 5 end in 0 or 5 so any number ending in one of these two digits will be divisible by 5. This was the basis for an assessment question in the 2004 national tests at Key Stage 2 which asked children for an explanation:

> John says,
> 'Every multiple of 5 ends in 5'
> Is he correct?
> Circle Yes or No   Yes/No
> Explain how you know
> . . . .
> . . . .

It was sufficient in this case to note that multiples of 5 could also end in zero.

For divisibility by 4 a number needs to be divisible by 2 and when the number is halved that half needs to be divisible by 2. For divisibility by 8 an even number can be halved and halved again. If the result is still whole then the original number is divisible by 8.

Numbers that are multiples of 9 have digit values that all add up to 9: for 18 add $1 + 8$, for 27 add $2 + 7$, for 36 add $3 + 6$ . . . for 108 add $1 + 0 + 8$, for 117 add $1 + 1 + 7$ . . . and this gives the divisibility test for 9. For multiples of 3 it is not quite so easy but the characteristic of all multiples of 3 is that the digits add up to 3 or 6 or 9 and this will guarantee that a number is divisible by 3. For divisibility by 6 a number needs to be even and a multiple of 3, so even numbers whose digits add up to 3 or 6 or 9 will work. The factor that is most difficult is the number 7 as there is no straightforward divisibility test for 7. Perhaps the reader may like to propose a divisibility test for 11 taking care that it applies to three-and four-digit numbers as well as two-digit numbers. There is a helpful article on divisibility tests by Tim Rowland on the nrich website (www.nrich.mathematics.org.uk).

It appears from results of national tests that children's experiences with factors and multiples is more limited than their experiences in learning 'tables facts' despite the fact that these experiences will give them more flexibility in working with numbers. A national assessment question at Key Stage 2 asked children to

> Write a 3-digit number which has 2 and 7 as factors.

For this question 77 per cent of pupils received no mark, a fact that gives cause for reflection on their learning needs.

Meaning of division

Most children in the early years of schooling are familiar with the idea of sharing and this is an important context for introducing division. Unfortunately, research has shown that too many children hold on to this naïve interpretation and they are inhibited in progressing to understand division more fully (Fischbein et al., 1985). Division starts with two distinct meanings, *sharing* (or *partition*) and *grouping* (or *quotition*) associated with different procedures and different verbal interpretations of the symbol. The two contexts: '24 apples shared among 4 children' and '24 apples put into bags of four' illustrate the distinction between the actions suggested although both will be symbolized as '24 ÷ 4'. The first suggests a sharing action while the second implies equal groups in a way that relates to repeated subtraction. The idea of making equal groups relates more easily to division as the inverse of multiplication and repeated subtraction as the inverse of repeated addition. Asking the question 'how many fours in 24?' then relates to the multiplication fact that '6 fours make 24'. This understanding of inverse operations is crucial for developing efficient ways for doing calculations. In the Netherlands, as part of the Realistic Mathematics Education approach (van den Heuvel-

Panhuizen, 2001), the idea of repeated subtraction is encouraged from an early stage rather than sharing, and this leads on to a written procedure that will be discussed in Chapter 5 that uses larger and larger 'chunks' to be subtracted.

At a later stage division will be associated with a ratio as the expression $3 \div 4$, for example, is identified with the fraction $\frac{3}{4}$ and can be re-interpreted as a quarter of 3, or 3 lots of $\frac{1}{4}$. This ratio notation is widely used when children reach secondary school and division is often presented as a ratio, such as, $\frac{24}{4}$ replacing the division symbol altogether. This introduces new opportunities for transforming the division by dividing or multiplying top and bottom by the same factor, in this case dividing both by 2 would result in the equivalent calculation related to $\frac{12}{2}$. Helping children to understand why this works could involve an illustration showing, for example, that 24 quarters is the same as 12 halves, or that $\frac{3}{4}$ of a shape is also $\frac{6}{8}$ when each section representing a quarter is halved, or is $\frac{9}{12}$ when each section is subdivided into 3 parts. Clearly it is not necessary for this particular calculation but could simplify considerably a calculation such as $\frac{240}{96}$ which can be reduced in stages to $\frac{5}{2}$.

Although this is a helpful way of simplifying division problems it is rarely considered in primary school. It can often replace the division algorithm and gives new opportunities when dividing by fractions or decimals as the numbers in the ratio can be adjusted to make the calculation easier. The calculation $25 \div 0.2$ can be seen as $\frac{25}{0.2}$ $\frac{250}{2}$ and the calculation $\frac{3}{4} \div \frac{1}{2}$ can be identified with $\frac{\frac{3}{4} \times 2}{\frac{1}{2} \times 2} = \frac{6/4}{1}$ giving meaning to fraction division by 'turning the divisor upside down and multiplying'.

The symbol '$\div$' is, however, important in primary school, and mental and written strategies for calculating will be identified in the next chapters.

Working with remainders

Many division calculations can be undertaken directly using the related multiplication fact but there are examples when the division is not exact and the idea of a *remainder* must be introduced. When the calculation is set in a context, as many test questions are, it will be important to determine what to do with the remainder as the following problem shows:

> How many 15-seater coaches are needed to take 432 children on an outing?

Research has shown that many children in England will ignore the context and give a numerical solution of '28 remainder 12' rather than making sense of the context and answering with '30 coaches' (Anghileri, 2001). The findings of this research contrast English children with Dutch children who are more used to context questions and better prepared to make a sensible solution. A similar question used in other research asking 'how many shelves of 4 metres can be cut from 3 planks of 10 metre length?' This also resulted in abstract calculations that took no account of the constraints implicit in the question. Where the first stage in problem solving involves imagining the problem, rather than going directly to an abstract calculation, children are more likely to be successful and errors will suggest that more experience is needed with real problem solving.

As suggested earlier in this chapter, the image of jumps on a number line can be helpful for both multiplication and division and shows well their relationship as inverse operations. A word of caution is needed where division is not exact and jumps on the number line will generate a remainder as in the calculation 17 ÷ 3. Jumping backwards in jumps of 3 from 17 will generate a complex pattern of numbers that are not familiar or related to multiples of 3. It is better to start at a number that is *close* to 17 but included in the pattern of multiples of 3 so that the whole number of jumps can be calculated first and then used to find the remainder. In this case 15 would be 5 lots of 3, so the remainder is shown as 2 units or ⅔ of a complete jump (Figure 3.5).

With problems set in a real world context numbers often do not divide exactly and 'remainders' need to be interpreted as fractions or as decimals. Working with calculators will enable children to calculate easily after selecting the appropriate operation and then to focus on interpreting the solution. Calculators may also provide an opportunity to help them understand the relationship between multiplication and division by using the inverse operation to check their calculations. Calculators can also be helpful in

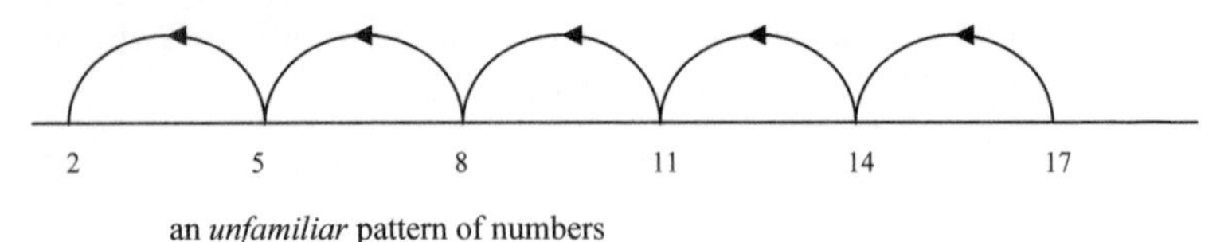

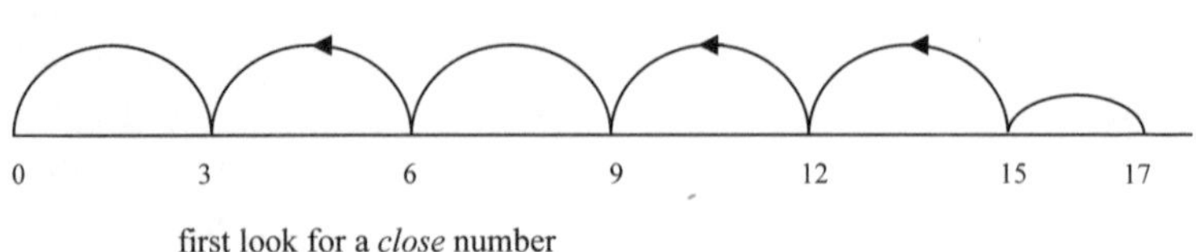

Figure 3.5 Empty number lines showing unfamiliar and familiar patterns of numbers

establishing the patterns of numbers that arise as remainders. Work with related division calculations can generate interesting solutions. The problems 'How many biscuits each if 26 biscuits are shared between 5 children, or 4 children, or 3 children?' leads to the very different results on a calculator that can be discussed and compared. Division by 4 and 5 will be easier to understand as the decimal fractions involved may be familiar. Division by 3 will introduce a long 'tail' of decimal numbers that may be related back to the original problem when looking for an explanation. Such difficulties should not be avoided but taken as a focus for discussion to raise children's awareness that these complexities arise.

Using spreadsheets to explore ratios set up between column entries can be a powerful way to develop experiences of multiplicative relationships that do not rely on repeated addition. This also reinforces the idea of the ratio between numbers in the different columns and this will be a most useful way of representing multiplication and division relationships in secondary schooling.

CHAPTER 4

Developing Additive Thinking

One of the biggest changes to mathematics teaching in recent years has been the emphasis placed on mental strategies and informal working, with delayed introduction of written methods for calculating. Government guidance stresses that in Key Stage 2 'pupils ... move from counting reliably to calculating fluently with all four number operations. They always try to tackle a problem with mental methods before using any other approach' (DfES, 2006c: 67). Much of the focus for teaching mathematics in the primary school is now on the development of strategic thinking and this continues into secondary school where pupils 'still need to practise and refine their mental calculation strategies' and the teacher's role is to 'guide pupils towards choosing and using methods' that are 'efficient and reliable' (DfEE, 2001: 11).

This focus on working mentally is in line with the needs of most adults in their working lives and reflects children's intuitive ways of working. It promotes independent thinking as appropriate strategies have to be matched to each problem rather than applying a 'blanket' algorithm for all calculations. But it can make teaching more difficult as the children's methods become the focus of attention rather than a 'correct' method taught by the teacher. Difficulties arise because calculation strategies are associated with different levels of conceptual thinking and individual children in the same class can show substantially different skills and understanding. A 'counting on' strategy for calculating 27 + 9, for example is clearly at a lower conceptual level than compensating using 27 + 10 − 1.

For previous generations, with the emphasis on written algorithms, the learning of mental strategies was rarely explicitly considered. Consequently, there is little established tradition in teaching mental methods or encouraging children's own informal and intuitive approaches. In this chapter, a number of approaches to calculating will be considered together with some analysis of the cognitive difficulties involved and the prerequisite skills that may be needed. Distinctions will be made between *sequential, partitioning* and *transformation* methods, and skills such as doubling and halving will be utilized. The sequential method (sometimes referred to as *complete number* method or *cumulative* method, for example by Thompson, 1997) is based on counting with whole numbers used throughout. One number is retained as a whole while the other is added or subtracted in (complete number) steps. In contrast, partitioning methods (sometimes referred to as *split number* methods) involve splitting up both numbers, usually into hundreds, tens and units and can involve adding or subtracting the digit values in the numbers usually arranged in columns (that is 23 + 54 is calculated as 2 + 5 and 3 + 4). Transformation methods will involve changing a calculation to a related calculation and then adjusting the answer.

Starting with counting

Until recently the crucial role of counting in developing mental calculation strategies has not been acknowledged. Rather, counting has been dismissed as an inefficient and time consuming way to work. It is true that the pupil who continues to count in ones to do a calculation such as 23 + 17 has many stages to master in developing efficiency, but it is inescapable that this is how calculating begins. In the past, counting was seen as a 'mechanical and meaningless activity, which should not be stimulated but instead should be replaced (as soon as possible) by the more meaningful place-value concept and its corresponding written (vertical) procedures for addition and subtraction, supported by structured apparatus' (Beishuizen and Anghileri, 1998). Researchers in the Netherlands, on the other hand, advocate an approach that does not suppress such informal activities as counting, but exploits them in ways that help children to develop more efficient ways of working without losing understanding of what the calculations mean (Buys, 2001). In Chapter 2, the activity of counting was associated with the imagery of numbers positioned on a number line with imagined jumps along that line in different directions associated with addition and subtraction. Research in the Netherlands has shown that this image of a number line can be a very effective support for working mentally and can provide an informal written method for more complex calculations (Beishuizen, 1999, 2001).

With this image of a number line, the calculation 4 + 3 can be represented with the image of a single 'jump' of 3 from the number 4 to the number 7 and this is the first conceptual step away from unitary counting (counting in ones). Together with combining sets of 4 objects and 3 objects, children come to recognize that 4 + 3 will always result in 7, which is an important *number fact* to be memorized. Such memorized facts will be the key to efficient calculating so time is well spent investing in a wide range of experiences that will help children to know these facts. It is also important to explore and discuss other relationships between 4 and 3 and 7 in order to develop flexibility in thinking about numbers. When it is recognized that '7 is 3 more than 4' or '4 is 3 less than 7' then addition and subtraction are related to the same number triple.

Recognizing that 7 is two 3s and one more will be important when division is introduced. This is helpful because children often remember the doubles facts earlier than any of the others. Talking about these different relationships, and later the way these ideas can be recorded, will set the foundation for many effective ways of calculating mentally. Children who are able to *recall* doubles of numbers like 25 and 15, and even doubles of 8, 16 and 32, will have many advantages when they come to more complex calculations. Associated with every double is a halving fact that can also provide an efficient shortcut in multiplication and division calculations.

The fundamental benchmark facts for addition and subtraction are the pairs that make ten: 1 and 9, 2 and 8, 3 and 7, 4 and 6, and the double: 5 and 5, and each pair works in the reverse order: 9 and 1, and so on. These facts are crucial for children to understand and memorize and the *hearts in love* (Menne, 2001) introduced in Chapter 3 seem to catch the attention and motivate the children. These hearts in love are not simply addition facts, but provide the information for related subtractions: 10 − 1 = 9; 10 − 9 = 1; 10 − 2 = 8; 10 − 8 = 2 . . . 10 − 5 = 5. These facts will be used extensively with numbers beyond ten as they relate to calculations such as 17 + 3, 27 + 3 . . . 70 + 30, 0.7 + 0.3 and form the basis for the method of *bridging through tens* discussed below. The empty number line provides an image associated with these facts as each can be seen as a jump from one number to another.

The empty number line and prerequisite skills for calculating mentally

The numbered line has been an aid to calculating for many generations but the innovation of removing all marks except those added by the individual to an empty number line discourages counting and favours flexible use of number facts (Beishuizen and Anghileri, 1998). These will be the basis for sequential calculating where numbers are built in a sequence of jumps. When calibrations are removed, initially leaving only the decade numbers and eventually leaving no markings, children begin to *imagine* and *visualize* for themselves the positions of numbers in a calculation. Individuals can draw their own line and it does not need to be straight as it is a personal aid to thinking rather than a prescribed written method. As children extend their range of counting numbers beyond 20, and start to recognize the pattern for constructing the symbols for bigger numbers, *locating* these on an empty number line will help provide imagery to support their thinking. The skill of locating numbers will later be related to ordering fractions and decimals and identifying distances between them (as indicated in Chapter 2).

When the counting sequence is known orally and related to the image on a number line it becomes possible to visualize the closeness of the different numbers. Ideas of *close numbers*, such as 19 being one less than 20, and 29 being one less than 30, will provide shortcuts in calculating and enable children to work at a higher conceptual level by transforming a calculation. The idea of locating numbers in relation to others will also be crucial when the idea of *rounding* is introduced. Rounding a number 'to the nearest ten', for example, is asking which decade number is closest and later rounding to one or two decimal places also asks which number is closest. Visualizing or illustrating the positions of numbers on a number line will help children develop this understanding of rounding.

Another skill that is vitally important for sequential mental calculation is the ability to *add 10 to any number*, or *subtract 10 from any number*, without having to count or calculate. Leading up to images on the empty number line, the bead string discussed in Chapter 2 presents a powerful image with contrasting colours of beads showing the structure of numbers and the visual effect of adding ten, for example 21 is two lots of ten and one more, and 31 is three lots of ten and one more. With a rigid bead bar a 'tens catcher', which is a small metal device (often home made) designed to move exactly ten beads at a time, can be used to investigate adding ten to a series of different numbers (Menne, 2001).

In order to reinforce the pattern of results for adding or subtracting ten, a calculator can be helpful so that individuals explore the patterns and then try to predict the resulting number and check to see if they are right. For children with different levels of understanding the start numbers can be adjusted. Counting on in tens from 27, for example, will be less challenging than starting from 327, or 897. Another resource that may be used is the hundred square which is normally a grid with the numbers 1 to 100 arranged in rows of ten or with the numbers 0 to 99. On this grid, counting, or adding one at a time, is associated with moving horizontally across the grid. When the end of one row is reached, the count continues at the start of the next row. Counting in tens is achieved by moving vertically down the columns. For subtraction the direction of movement is reversed and tens are subtracted by moving vertically up a column. This resource has the advantage that it is more compact than a number line extending to 100 but research in the Netherlands has shown that lower-achieving children can become very confused with the complex moves that are necessary and that they gain better understanding using an empty number line. Once again, the power of the number line emerges as children visualize where the numbers are and 'see' jumps of different sizes (Anghileri, 2006b).

Chunking numbers

In preparation for addition and subtraction there has been a long tradition of teaching 'place value', splitting numbers into tens and units, and children are given a lot of experience with questions such as 'how many tens in 57 and how many units?'. Such understanding forms the basis for partitioning methods of calculating. This preparation was important for the column arithmetic taught to previous generations but children today need more flexibility and using only tens and units can inhibit some efficient methods for calculating. If 57 were to be divided by 3, for example, there would be advantages in splitting 57 into 30 and 27, rather than 50 and 7. Even place value needs to be used flexibly as $1256 \div 6$ is a much more difficult calculation when considered as $1000 \div 6$, $200 \div 6$, $50 \div 6$ and $6 \div 6$ than when it is considered as $1200 \div 6$ and $56 \div 6$.

It is certainly important that children are aware of the tens and ones structure of numbers and their ability to extend a counting sequence will display this understanding, but asking 'how many tens in 57?' is not the same as accessing their understanding that 57 is 50 and 7, and seems a pointless activity to some children. A '5 in the tens column' is a much more abstract idea than 50 which is a whole entity in its own right. Although verbal counting gives many clues to the ways numbers can be split, it is important to give children practical experiences with a variety of materials from base ten apparatus to colouring grids and using money. The latter is a particularly rich resource when 20p coins and 50p coins are used as well as 10p and 1p coins. A classroom shop, collecting for charity or a 'bring and buy' sale can all give motivation for using coins and learning to make different amounts. A board game using a dice or spinner showing these and other different values can be used in games where addition and subtraction are required to move around the board. The idea of moving forward and backward also reinforces the idea of addition and subtraction being *inverse operations.* A rich resource for games and puzzles is the nrich website, www.nrich.maths.org which accompanies a great wealth of problems with notes, hints, solutions and printable pages. An example from this website is the 'Chocoholics problem': George and Jim want to buy a chocolate bar. George needs 2p more and Jim needs 50p more to buy it. When they put their money together, it is still not enough to pay for the chocolate bar. How much is the chocolate bar?

Mental methods and informal written methods for addition

Researchers have analysed children's own mental methods and informal written methods making distinctions between complete number methods and split number methods. In research that analysed APU (Assessment of Performance Unit) data from 1987, Foxman and Beishuizen (1999) acknowledge that neither method will have been taught at that time but that each represents an untaught intuitive approach. In today's classrooms these methods should appear as children's own suggestions to be shared and refined through discussion. This research suggested that complete number methods are more successful than split number methods for more complex calculations, particularly subtraction.

Complete number methods

In complete number (sometimes called sequential) methods one of the numbers is used as the starting number and the other is added in steps, for example 25 + 67 may be calculated as follows:

25 + 60 = 85 and 85 + 7 = 92
or
67 + 20 = 87 and 87 + 5 = 92

Care must be taken in recording these steps as the following is mathematically incorrect and needs to be discouraged:

25 + 60 = 85 + 7 = 92 **mathematically incorrect!**

These examples already show some sophistication and efficiency using number facts to add the final units and could be broken down into easier steps using number bonds for ten:

25 + 67 25 + 60 = 85 and 85 + 7 = 85 + 5 + 2 = 92

This method is sometimes called *bridging through tens* and has the advantage that known facts like 85 + 5 are used rather than the more difficult 85 + 7.

At a lower conceptual level a more naïve method for adding 60 to 25 is to use steps of 10 and count on in ones the additional 7. This will happen where children are not comfortable to connect

20 + 60 to 2 + 6 and more experience will be needed in making these links.

The above methods appear clumsy when recorded horizontally in this way but show the steps done mentally and help children keep track of a calculation as they are doing it. A more efficient *jotting* can be illustrated on an empty number line (Figure 4.1).

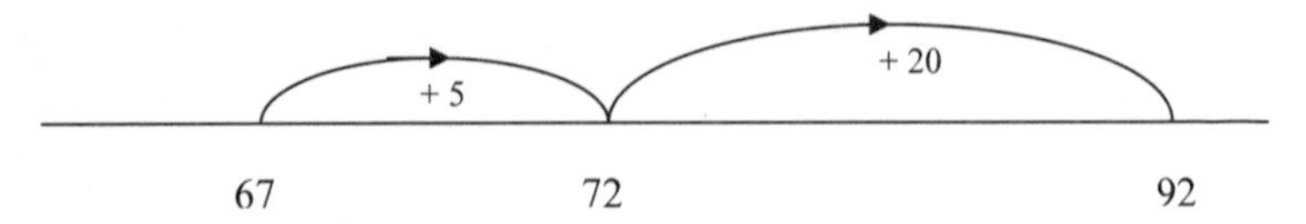

Figure 4.1 67 + 25 illustrated on an empty number line

Here the empty number line provides a personal record to support mental thinking. It will also be an efficient written method for more complex calculations as examples later in the chapter will show. This variation in strategy makes a good basis for discussion and the plenary in a lesson can be used for children to compare their different approaches and learn from each other. The very act of describing your method, aided by the visual representation, helps children reflect on their methods and gain a deeper understanding.

These examples show the element of choice that characterizes complete number methods as each individual will choose the starting number and the steps to use in adding on the second. Through discussion children who are using inefficient methods can be made aware of ways to progressively gain efficiency without losing ownership of their method.

Split number methods

In contrast to complete number methods, in split number (sometimes called partition) methods both numbers are split into their tens and units values, which are added independently to give partial sums, usually starting with the larger parts. The same addition, 25 + 67 is calculated:

$$20 + 60 + 5 + 7 = 80 + 12 = 92$$

This can be done mentally and recorded horizontally and it differs from the traditional algorithm where smaller values are added first in preparation for written recording in columns. Here the term split implies the use of tens and units and will be contrasted with the term chunking met in Chapter 2, where the parts that are used will depend on the calculation in question. This splitting method is not so easily illustrated and requires some re-organization of the partial results which both need to be held in memory when the calculation is done mentally. There is also no element of choice and where the teacher focuses too much on this method it can become procedural. That is, children will try to replicate what they have been shown rather than thinking through their own method.

For many adults the *traditional algorithm* will form the basis for calculating mentally 25 + 67 and will start with addition of the units followed by addition of the tens:

> 5 + 7 = 12 and 2 + 6 = 8 resulting in 2 units and 8 + 1 = 9 tens

This method relates closely to column methods for written calculations where calculating the units first means that the tens total is adjusted before it is recorded. Although this method may seem 'natural' to many adults this is probably due to the extensive practice of column arithmetic that used to typify the primary mathematics curriculum. As a mental method, addition of the largest parts first has the advantage that these indicate a better approximation to the solution that is calculated while there is more 'free memory'.

Transforming methods

There will be no single 'correct' way to do addition calculations but individual children will choose methods that are in keeping with their own thinking. Take the problem 'I have 25p and my friend had 28p. How much do we have together?' which relates to the symbolic expression 25 + 28. These numbers may trigger a known double 25 + 25 = 50, particularly in the context of money, and the total of 53p is achieved very efficiently. Some children will progress to more efficiency with the following methods for calculating 25 + 68:

> 25 + 68 = 30 + 63 *transforming* the calculation to an equivalent one
> 25 + 68 = 25 + 70 – 2 *compensating* by adding too much and subtracting the excess

These are examples of methods known as *clever calculating* which can be encouraged for those children who are efficient at calculating using more straightforward methods.

The whole idea of *developing number sense* is for individuals to have confidence to tackle calculations choosing efficient methods that they understand.

Choice of appropriate methods is required for national assessment questions which rarely give a straightforward addition calculation. The national test question from Key Stage 2 in 2004

$$\square + 85 = 200$$

suggests that number sense could enable this calculation to be connected with $\square + 85 = 100$ and the image of a jump from 85 to 100 of 15 to be visualized or drawn on an empty number line.

The following illustration from the Key Stage 3 Framework (Figure 4.2) shows addition of decimals illustrated on an empty number line using the compensation method of adding too much and then subtracting the excess.

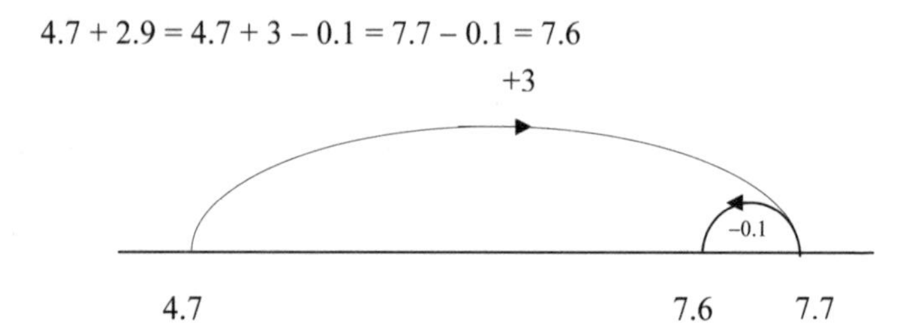

Figure 4.2 A compensation method for calculating 4.7 + 2.9

Each of the methods illustrated above shows how a written record can support mental thinking by reflecting the procedures used in the solution. In some cases, particularly where there are several numbers to be added, setting them in a vertical arrangement can help but careful attention needs to be paid when the numbers do not all have the same number of digits. Adding 312 + 12.2 + 57 will lead to errors for children who have not had enough experience of arranging their own calculations. For previous generations a vertical arrangement was used with a rigid procedure 'starting with the units, carrying tens' and so on. As discussed in Chapter 1, it was not important for children to understand this algorithm but to practise it through many exercises. This is no longer the case as curriculum guidance for Scotland indicates: 'The purpose of teaching algorithms for calculating (step by step procedures for adding, multiplying, etc.) is to help pupils to understand number and use it effectively. Written algorithms are only one possibility. Mental algorithms are often more powerful; they are much more varied and need to be recognised, encouraged and discussed. Algorithms can also be devised for the calculator'. (Learning and Teaching Scotland, 1991)

Mental and informal written methods for subtraction

When children are engaged in problem solving they will meet addition and subtraction in contexts where they will have the opportunity to see them as inverse operations. As discussed in Chapter 3, each can be related to counting forwards or backwards with images of jumps in different directions on the number line. In the same way that split number methods and complete number methods were contrasted as mental methods for addition they can be used for subtraction calculations. As children find working 'backwards' in subtraction more complex than addition, another method of *complementary addition* will be discussed as well as *clever calculating.*

Complete number methods

Using a complete number method a subtraction calculation is associated with counting backwards in steps with growing efficiency. For the example 63 – 27 it is possible but very inefficient to count back 27 in ones from the number 63 and children who attempt this are rarely successful. Much more efficient is to subtract 20 and then count back the remaining 7 from 43

$$63 - 20 = 43 \text{ and } 43 - 1 - 1 - 1 - 1 - 1 - 1 - 1 = 36$$

This could be made more efficient using the idea of bridging through tens using known facts $43 - 3 = 40$ and $40 - 4 = 36$. This would mean that 7 could be subtracted in two steps of 3 and then 4. Children will need support in deciding how to record this calculation and the following is only one possibility.

$$63 - 20 = 43 \text{ and } 43 - 3 - 4 = 40 - 4 = 36$$

An alternative complete number method would be for the 7 to be subtracted first, bridging through 60 if necessary and then subtracting 20:

$$63 - 3 - 4 = 56 \text{ and } 56 - 20 = 36$$

Using the empty number line a complete number method could involve finding the steps needed to get from 63 back to 27 as subtraction is interpreted as 'the difference between' 63 and 27. When subtraction is in context, for example the difference between two ages or two measurements, the choice of strategy may be influenced by the contextual images (Figure 4.3).

These different methods will result in different illustrations and written recording but can be equally efficient and leave choice to the individual. This element of choice again suggests that individuals take ownership of their preferred method rather than replicating one used by their teacher. Discussion can be fruitful as through:

> explanation and discussion of their own and other people's methods (pupils) begin to acquire a repertoire of mental calculation strategies. It can be hard to hold all the intermediate steps of a calculation in the head and so informal pencil and paper notes, recording some or all of their solution, become part of a mental strategy. These personal jottings may not be easy for someone else to follow but they are an important staging post to getting the right answer and acquiring fluency in mental calculation.
>
> (DfEE, 2001: 10)

For some questions, children encountered difficulties in determining a correct model to solve the question. One example was question 32, where children were asked *How much taller is Sunil than Megan?* The following example demonstrates an efficient model that one child used to solve the question.

Figure 4.3 An example where the context may influence the calculation strategy (QCA, 2003b)

Transforming methods

Using the connection between addition and subtraction it is possible to avoid working 'backwards' for subtraction and transform each calculation to an addition in a method called complementary addition. This will not be easy for children who always interpret subtraction as 'take away' but relates well to the idea of the 'difference between' two numbers and is clearly illustrated on an empty number line.

The calculation 25 – 17 may use jumps backwards from 25 of 10 and then 5 and then 2 to reach the answer 8 (Figure 4.4).

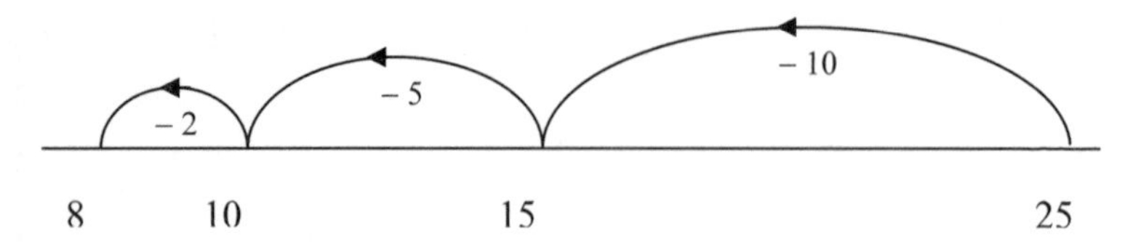

Figure 4.4 Calculating 25 – 17 by jumping back on an empty number line

Alternatively this calculation may be identified as the difference between 25 and 17 (or 'what do I need to add on to 17 to get to 25?') and illustrated with jumps forward. Now the skill of *locating* numbers plays an important role as the difference is seen as a jump of 3 to 20 followed by a jump of 5 to 25. It must be noted that the image this presents is very different to the subtraction above and the answer is given by the total size of the jumps and not the numbers reached (Figure 4.5).

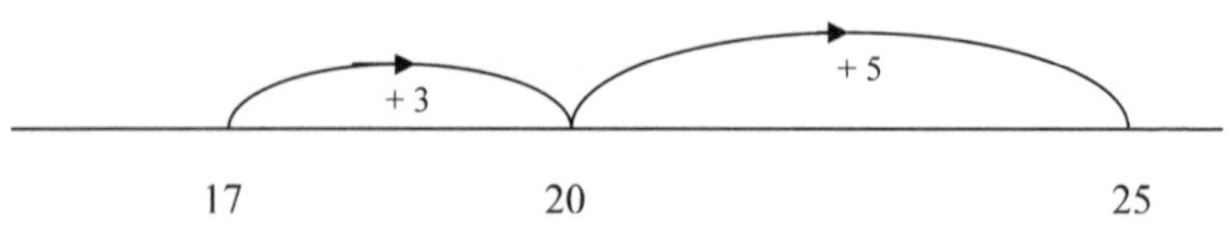

Figure 4.5 Calculating 25 – 17 by jumping forward on an empty number line

For problems involving decimals, for example 5.4 – 1.8, this method of subtracting by adding on to the smaller number, although conceptually more complex, will often result in easier calculating (Figure 4.6).

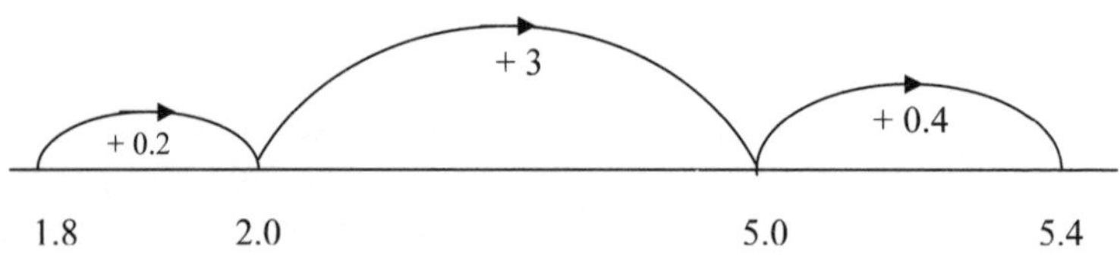

Figure 4.6 Calculating 5.4 – 1.8 using complementary addition on an empty number line

Another highly efficient transforming strategy is the method of compensating. For subtraction this method involves subtracting too many, using a known fact, and then adding to balance the calculation. With numbers close to decade numbers, such as 37 – 19 or 373 – 198, or close to whole numbers, such as 3.5 – 1.9 or 8 – 5 $\frac{7}{8}$, it is an effective strategy and for decimals or fractions this can reduce the need for complex subtractions as the following example shows (Figure 4.7).

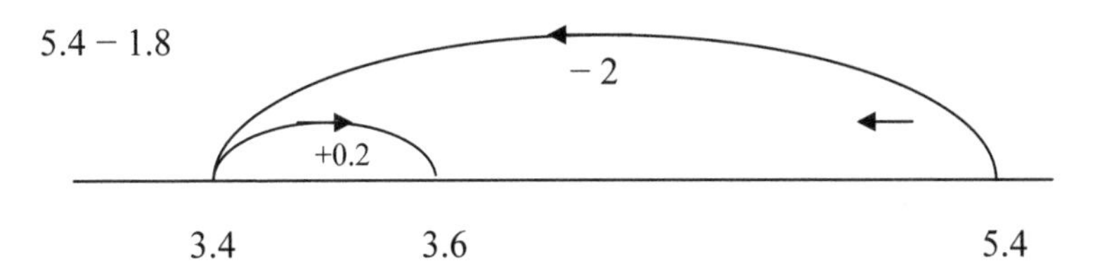

Figure 4.7 Calculating 5.4 – 1.8 using a compensation method on an empty number line

Split numbers methods

For some subtraction calculations involving two- or three-digit numbers a split numbers method can be straightforward but complexities arise when a digit in the number to be subtracted (the *minuend*) is greater than the associated digit in the number being subtracted (the *subtrahend*). In the example 68 – 23 the numbers can be *partitioned* into tens and units which are subtracted separately:

60 – 20 = 40   8 – 3 = 5 and 40 + 5 = 45

If the example is 63 – 28 then this method needs considerable adjustment as 8 'cannot' be subtracted from 3 (unless negative numbers are introduced). Where column subtraction is introduced it is a common error for children to take the smaller number from the larger as the example from a national test report (QCA, 2004a) suggests:

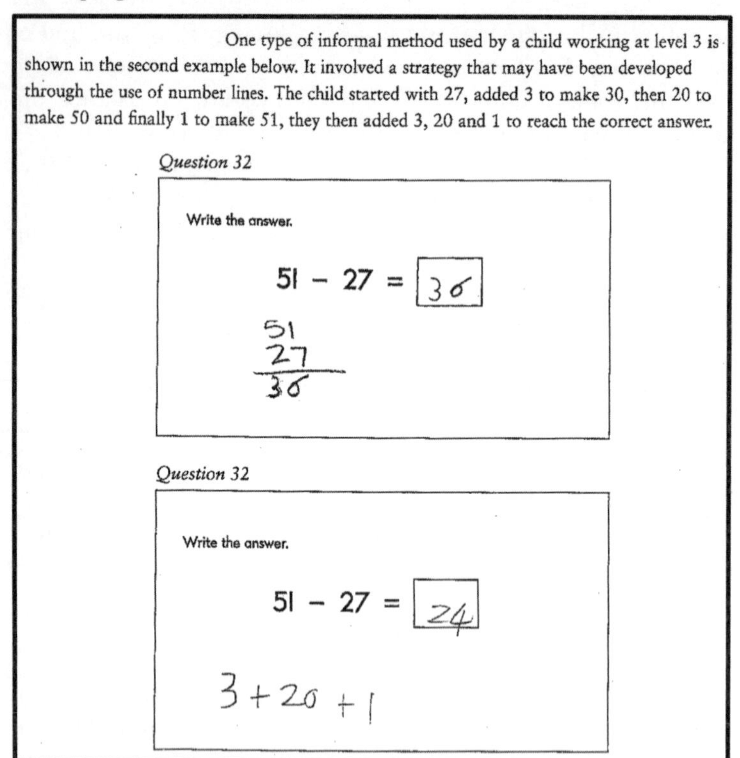

One type of informal method used by a child working at level 3 is shown in the second example below. It involved a strategy that may have been developed through the use of number lines. The child started with 27, added 3 to make 30, then 20 to make 50 and finally 1 to make 51, they then added 3, 20 and 1 to reach the correct answer.

Question 32

Write the answer.

51 − 27 = 36

51
27
36

Question 32

Write the answer.

51 − 27 = 24

3 + 20 + 1

Figure 4.8 A common error in the column subtraction method

This report illustrates an effective strategy using an empty number line. Some individuals will overcome this type of difficulty by introducing negative numbers

$$63 - 28 = (60 - 20) + (3 - 8) = 40 + (-5) = 35$$

This is sophisticated thinking and should be encouraged but not expected from the majority of children in a class. The types of number in a subtraction calculation will affect the complexity of the method used and children need support in making decisions about which method will be most effective for them to use. Making good choices is an important skill for children to develop.

Progression in written methods

For previous generations the standard algorithms using numbers arranged in columns have been the pinnacle of achievement in addition and subtraction calculations because they provide a concise and ordered recording. Until recently traditional algorithms have been taught without question and with some suggestion that they always work. Unfortunately, the reality is that some children become confused by the complex procedures involved (for example, when subtraction involves taking a larger digit from a smaller one) and the resulting calculations involve errors. What is needed is a continuous progression from mental methods described above that are built on number sense, to an organized record of the calculation process. This may involve the arrangement of numbers in columns or may involve an empty number line as the most efficient written record.

Following DfES guidance

In their guidance on calculations DfES describe in some detail stages in developing a written record for addition and for subtraction (DfES, 2006c) as the following extracts will show (Figures 4.9 to 4.12). As indicated throughout this chapter it is important for children to keep hold of the understanding of numbers and number operations that they are developing and make choices of the written method they feel most appropriate. Sometimes this may be the column method but often there are opportunities for alternative strategies based on the numbers involved and that require less or easier working.

It is not clear why so much emphasis remains on column methods as in many calculations the arrangement in columns will be no more efficient than a horizontal format, particularly where partial sums are calculated mentally. The national test question 309 – 198 illustrated earlier in Figure 3.1 was such an example that was efficiently recorded using an empty number line and a column method would have been no more efficient. There is a danger that any standard method will be valued by children more highly than their own individual thinking because it has been taught by the teacher with an expectation it should be used. Standard methods

Identified as Stage 1 is 'Using the empty number line' with steps often bridging through a decade number:

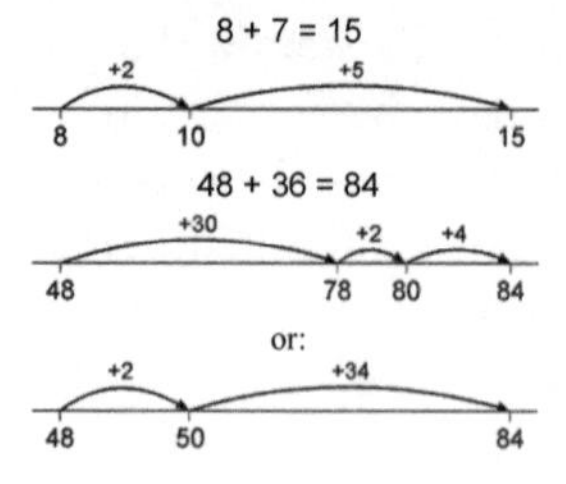

Stage 2 is 'Partitioning' with a horizontal written record and then vertical recording:

$$47 + 76 = 47 + 70 + 6 = 117 + 6 = 123$$
$$47 + 76 = 40 + 70 + 7 + 6 = 110 + 13 = 123$$

Partitioned numbers are then written under one another:

```
  47  =  40 +  7
+ 76     70 +  6
        110 + 13  = 123
```

Figure 4.9 Guidance for written methods for addition: stages 1 and 2

Stage 3 involves the 'Expanded method in columns':

Adding the tens first:

```
   47
+  76
  110
   13
  123
```

Adding the ones first:

```
   47
+  76
   13
  110
  123
```

Finally Stage 4 identified for addition is the traditional algorithm complete with numbers 'carried' and recorded under the line (sometimes referred to as 'on the door step' to aid memory)

```
   47      258      366
+  76    +  87    + 458
  123      345      824
  1 1      1 1      1 1
```

Figure 4.10 Guidance for written methods for addition: stages 3 and 4

As for addition, stage 1 in subtraction is use of 'The empty number line' but already introducing a column record:

74 – 27 = 47 worked by counting back:

−3 −4 −20
47 50 54 74

The steps may be recorded in a different order:

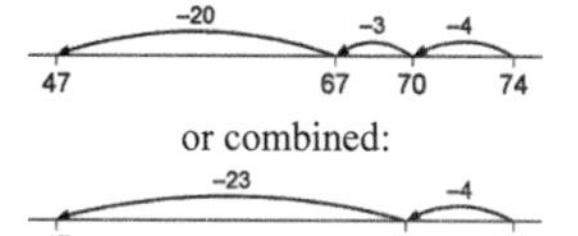

or combined:

Then with counting up rather than counting back:

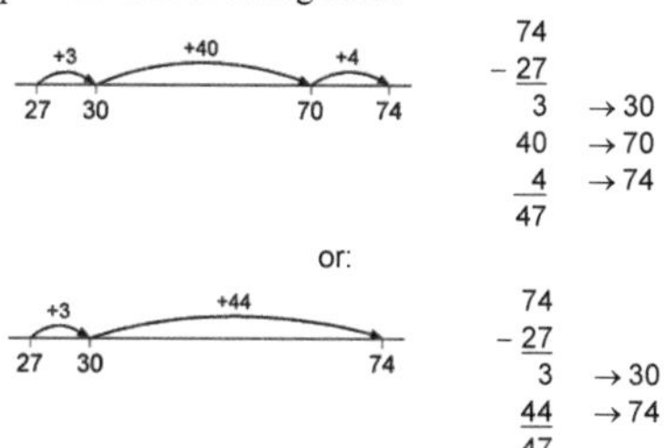


```
  74
- 27
   3  → 30
  40  → 70
   4  → 74
  47
```

or:

```
  74
- 27
   3  → 30
  44  → 74
  47
```

Figure 4.11 Guidance for written methods for subtraction: stage 1

Stage 2 is identified as 'Partitioning' with a horizontal record:

74 – 27 = 74 – 20 – 7 = 54 – 7 = 47

74 – 27 = 70 + 4 – 20 – 7 = 60 + 14 – 20 – 7 = 40 + 7

This method of recording links to counting back on the number line and a jotting with 'bridging through a decade number' may be more appropriate.

−3 −4 −20
47 50 54 74

Stage 3 is an 'Expanded layout, leading to column method' introducing ideas of *decomposition* of the top number:

Example: 74 – 27

```
  70 + 4        60    14        6 14
- 20 + 7        70 +  4         7 4
              - 20 +  7       - 2 7
                40 +  7         4 7
```

Example: 741 – 367

```
  700 + 40 + 1      600   130   11      6 13 11
- 300 + 60 + 7      700 + 40  + 1       7  4  1
                  - 300 + 60  + 7     - 3  6  7
                    300 + 70  + 4       3  7  4
```

Figure 4.12 Guidance for written methods for subtraction: stages 2 and 3

may also be used procedurally in an unthinking way and for all numbers and may inhibit clever calculating. It is important that written recording is developed progressively to support children's thinking and that individuals retain ownership of methods they understand rather than replace these with a standard method.

These examples show clearly the complexities that column subtraction can introduce that may not exist where the horizontal format is used. The reader might take a few minutes to consider alternative ways for recording the final calculation above 741 – 367. What has been lost in this formalization of a written record is the idea of complementary addition as an effective and efficient method for subtraction that will avoid the need for decomposition.

For many calculations the column method is not the most efficient, for example, 6067 – 5970 seen in Chapter 1. It is also possible that teaching traditional column methods will not encourage the strategic thinking that leads to clever calculating such as transforming a calculation to make it easier. It is, however, important that children organize their recording and this needs to be done without inhibiting their choice of strategy.

Developing additive thinking

It has been important to take a detailed look at progression in teaching written methods for addition and subtraction as these continue to be a central focus in many classrooms. It is important, however, to note that these methods do not constitute number sense and can inhibit children in being flexible problem solvers. In Chapter 8 of this book all the examples will demonstrate that using number sense in problem solving will not depend on careful application of a standard procedure but will more often require understanding of the operations and their relationships.

CHAPTER 5

Developing Multiplicative Thinking

For addition and subtraction the most important 'benchmark facts' underpinning additive thinking are the number bonds of ten and the 'benchmark patterns' that arise when 10 is added to, or subtracted from, any number. In multiplication there are many more facts, starting with the 'tables facts', that will underpin multiplicative thinking, and 'learning by heart' is only one strategy for coping. Although it will go only part way to explaining multiplication, chanting and memorizing the counting pattern for multiples will build familiarity with numbers that occur frequently, such as 36, and those that never appear, such as 37. This adds to children's number sense and when the different patterns are discussed this will help them to develop a 'feel' for numbers in a multiplicative situation. In Chapter 3 it was suggested that some of the facts that are difficult to memorize, such as 8×7, can be more easily derived from known ones. Doubling $4 \times 7 = 28$ would be a good way of deriving this fact, or adding another 7 to $7 \times 7 = 49$. Rapid recall is the most efficient provided it is achieved in conjunction with, and not at the expense of, strategic thinking. Using games and puzzles will give a purpose to generating multiplication facts and a calculator can be an effective support during the learning stage.

A calculator will also give access to multiplication and division facts well beyond the times tables and can be helpful in giving children a sense of the way numbers are related. It is always helpful to use a calculator in a thinking way and interventions by the teacher can provoke this thinking. The activity of using a four-function calculator to generate the multiples of 3 (by pressing '3', '+', '+' and then the '=' button repeatedly) can be accompanied by the question 'Will the number 100 appear in this pattern?' and then reasons can be discussed. Another example can occur when they are exploring with the '÷' button on a calculator and need to interpret a decimal number output. This can be a very early experience as children experiment and input expressions like '5', '÷', '2'. If this is identified with the meaning of 5 ÷ 2 as '5 shared between 2' or 'how many 2s in 5?' the output of 2.5 will be associated with other experiences of division and may be interpreted as 2½. Number sense develops as an eclectic collection of facts and the identification of 0.5 as a half is a piece of information that will only gradually be integrated within an understanding of decimals. A more systematic exploration of division on a calculator, for example investigating 10 ÷ 2, 9 ÷ 2, 8 ÷ 2, . . . 4 ÷ 2, 3 ÷ 2, 2 ÷ 2, 1 ÷ 2, can provide the basis for a whole class discussion involving decimal numbers. Many patterns are worthy of exploration with a calculator and dividing numbers by 10, starting with single-digit numbers, gives further meaning to tenths.

Deriving new facts from known facts

This idea of finding new facts from old can be very effective for showing connections among multiplication and division facts and simplifying some calculations. As well as the commutative rule discussed in Chapter 3 that relates facts in all the multiplication tables there are many ways that connections can be made.

Making connections

If 8 × 7 = 56 is known, for example, then 80 × 7 and 8 × 70 can quickly be found as 8 × 7 × 10 using the associative rule and resulting in 560. Sometimes this pattern of multiplying by ten is described as 'just add a nought' but this is a poor description as it will not work in all cases. Take, for example 1.2 × 10, which does not result in 1.20 although this will become a common error where children are used to 'adding a nought'. Instead it is necessary to talk about digits moving to the left of the decimal point as units become tens, tenths become units and so on. In this particular example, 1.2 × 10, it could help understanding if it is taken in two parts as 1 × 10 and 0.2 × 10 as this would give a good first indication of the size of the answer. A web diagram can be useful for deriving many new facts (Figure 5.1) from a given starting point and individuals can work with ideas appropriate for their own level of cognitive challenge. Some may choose to use a calculator to derive related facts and the recording they make will tell the teacher a lot about their thinking and their level of understanding. When used as a class activity, producing these web diagrams gives children the opportunity to explain their thinking and for them to see the ideas of others.

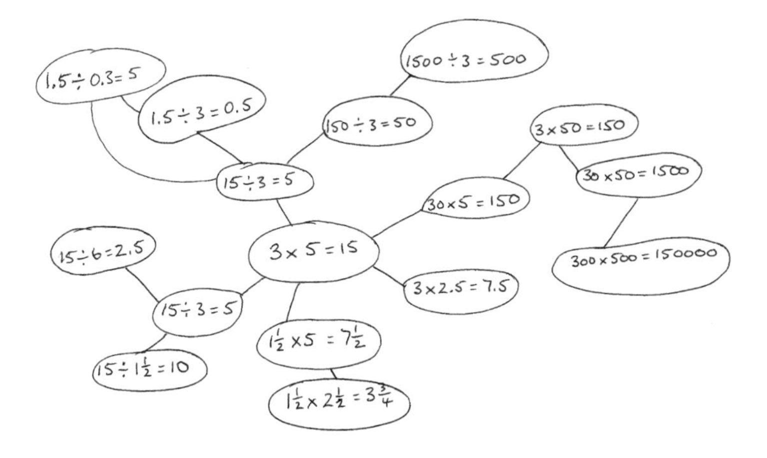

Figure 5.1 A web diagram of multiplication and division connected facts

Doubling and halving

The web diagram above shows new facts derived from known facts by doubling and halving and this provides an effective strategy when children understand the effects on multiplication and division. For multiplication, bigger products can be derived, for example from $4 \times 7 = 28$, doubling either of the numbers will double the product, $8 \times 7 = 4 \times 14 = 56$. This becomes clear if the image of an array, or a rectangle on squared paper, is used and one of the dimensions is doubled. This also provides a strategy for multiplying any number by 4 or by 8 (and even by 16, 32 and so on) by repeated doubling as the second calculation in Figure 5.2 illustrates:

The following responses demonstrate a variety of successful strategies used by some children in solving this multiplication calculation.

This child used a good strategy of partitioning to solve the problem.

Calculate 2307 × 8

2000 × 8 = 16000
300 × 8 = 2400
7 × 8 = 56
18456

18456

Test A, Q13, example of a correct response.

The second response demonstrates a successful informal method of doubling 2307 three times.

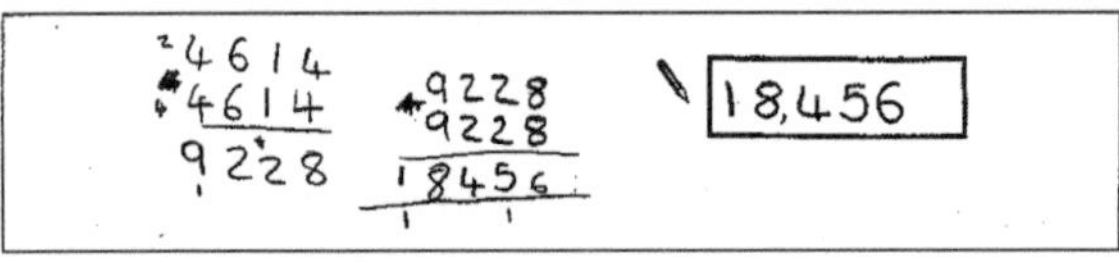

Test A, Q13, example of a correct response.

Another successful strategy used by a number of children was the grid method.

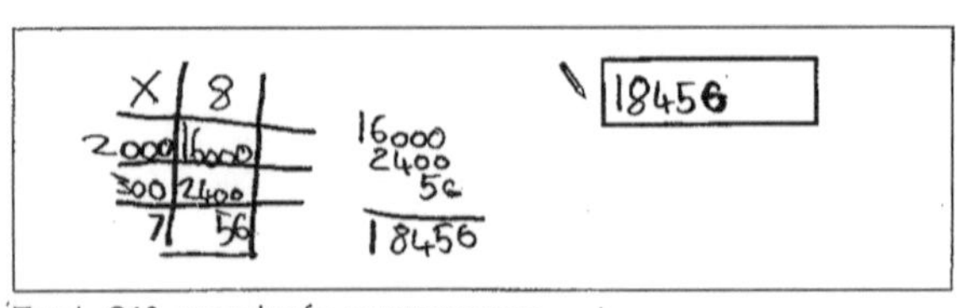

Test A, Q13, example of a correct response.

Figure 5.2 Different methods for calculating 2307 × 8

This illustration of children's methods used for Key Stage 2 assessment questions (QCA, 2003a) shows effective methods including partitioning, repeated doubling and the grid method which will be discussed later in this chapter.

In a similar way to doubling, multiplying one of the factors by any number will have the effect of multiplying the total (the *product*) by that number. The result of multiplying 40 × 3 will give ten times the total of 4 × 3 = 12 and 400 × 3 will give one hundred times 12. When both numbers are multiplied by ten as in 30 × 40 this will give one hundred times the product as 30 × 40 = 3 × 10 × 4 × 10 = 3 × 4 × 10 × 10. Again, this can be illustrated by increasing both of the dimensions of an array by the factor 10 and children will understand better if they have a visual image to associate with this idea. This will provide a meaningful strategy for calculating with decimals as both 0.4×3 and 4×0.3 will give one tenth of 4×3, and 0.4×0.3 will give one hundredth of 4×3, that is, 12 hundredths or 0.12. This will be discussed further in Chapter 7.

Multiplication by 5 gives another example where halving can lead to an effective mental calculation method as the result will be half of the product when multiplying by 10. A calculation such as 68×5 can be found by halving $68 \times 10 = 680$ to give 340 which presents less working and requires less to be held in memory than 60×5 added to 8×5.

In this section we have seen doubling can lead to effective ways of multiplying by 2, 4, 8 and 5. It is a good strategy for children to start multiplication calculations (and division as will be seen later in this chapter) by writing down the results of multiplying $1\times$, $2\times$, $4\times$, $8\times$ and $10\times$. From these they can derive new facts by adding relevant subtotals: $6\times$ will be $2\times$ added to $4\times$, $9\times$ is $10\times$ minus $1\times$, and many others can be found in this way.

Using factors

A multiplication calculation can be made more accessible for a mental method if one of the numbers is split into factors 24×4 as $8 \times 3 \times 4$ that is '24 lots of 4' is the same as '3 lots of 8×4' or '8 lots of 3×4' or '12 lots of 2×4' or '2 lots of 12×4'. In this way choices can be made about which known facts will lead to the easiest calculations and doubling can be incorporated. Many calculations involving larger numbers can be broken down in this way as children's sense of factors and multiples continues to develop. Some numbers will crop up more frequently than others as they have more familiar factors. Numbers such as 75, or 96 will come to trigger a better feeling than 79 or 97 which have less familiar factors. Assessment questions are sometimes based on these friendly numbers and recognition of their factors can provide an effective strategy for calculating.

A national test question at Key Stage 2 from 1998 required explicit identification of multiples and remainders:

Circle **one number** on the grid which can be **divided by 9** with a **remainder of 1**.

97	98	99
107	108	109
117	118	119

Knowing the divisibility test for nine, numbers having digits that add to nine (see Chapter 3), would help identify 99, 108 and 117 as multiples of nine and then choosing the next one in the sequence of numbers would be needed for a remainder of one.

Partitions and chunking for multiplication

Figure 5.2 earlier in this chapter shows a method for calculating 2307 × 8 by partitioning or splitting the number 2307 and doing the calculation in parts (2000 × 8) + (300 × 8) + (7 × 8). In this example the working is well organized and the arrangement of the subtotals shows good understanding of the importance of place value. The third illustration is effectively the same but is organized in a grid which can help children to structure a written record of their calculation and it also relates to the area model for multiplication.

Central to this way of working with larger numbers is an understanding of the ***distributive rule*** (a third ***law of arithmetic*** along with the commutative rule and the associative rule introduced in Chapter 3) that will be used when numbers are partitioned. The distributive rule enables numbers to be split so that 24 × 3 = (20 + 4) × 3 = (20 × 3) + (4 × 3) [the additive part is 'distributed' over the multiplication]. A number can be chunked in different ways to make use of facts that are known, for example 7 × 8 = (5 × 8) + (2 × 8).

This rule makes sense when it is put into words as, for example, '24 lots of 3' will be the same as '20 lots of 3' added to '4 lots of 3' and this can be illustrated on squared paper using the area model for multiplication with 24 columns and 3 rows partitioned into parts (Figure 5.3).

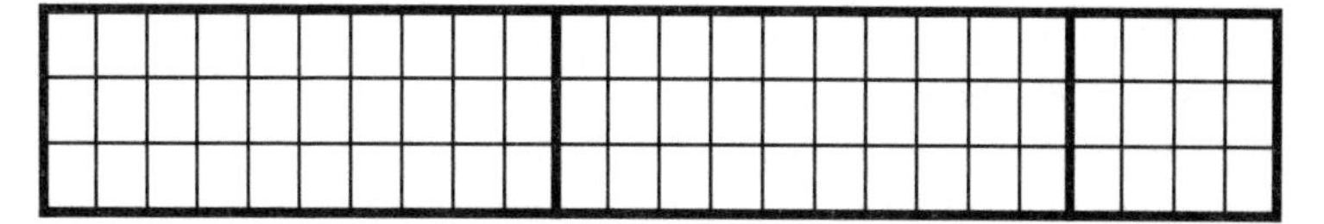

Figure 5.3 An area model for 24 × 3

As children become familiar with the fact that 20 × 3 = 60 this illustration can be contracted and provides a grid arrangement for recording multiplication. This grid method can later be extended for multiplying two- and three-digit numbers (Figure 5.4).

	3
20	60
4	12

So 24 × 3 = 60 + 12 = 72

	30	**2**
20	600	40
4	120	8

So 24 × 32 = 600 + 40 + 120 + 8 = 768

Figure 5.4 The grid method for multiplying two-digit numbers

This is sometimes referred to as an 'informal' written method despite the standardization of the format. For some children it will provide the most effective written method as it builds progressively from their intuitive understanding.

Splitting a number is a good way to take a multiplication in steps but the partition does not need to involve hundreds, tens and units. For the calculation 26 × 3, for example, the known fact that 25 × 3 = 75 could be used, adding one more 3 to get the total 78. This idea is called chunking because partitions are usually associated with splitting numbers using place value. (An alternative way to calculate 24 × 3 could also use the known fact 25 × 3 with 3 subtracted from the resulting 75.) The multiplier may also be split into chunks, for example 32 × 6 can incorporate multiplying 32 by 3 and then doubling the answer 96 which may be more accessible to some children than multiplication by 6.

Each multiplication calculation may invite different shortcuts and routes to efficiency according to the numbers involved and children with number sense will often end up doing easier calculations than those who always use standard procedures.

Children who retain ownership of their own methods rather than trying to replicate a standard method taught by the teacher can often be more resourceful in finding ways to simplify a calculation. The illustration from a report on national tests in 2001 (Figure 5.5) shows a child who has confidence to do 509 × 24 as '10 lots of 509 and 10 lots of 509 and 4 lots of 509' added together to get the correct result where the standard method may have been too complex to understand.

Other children used methods based upon partitioning of the numbers; such methods were seen more often among children working at levels 3 and 4. An example of such a method used correctly is shown below.

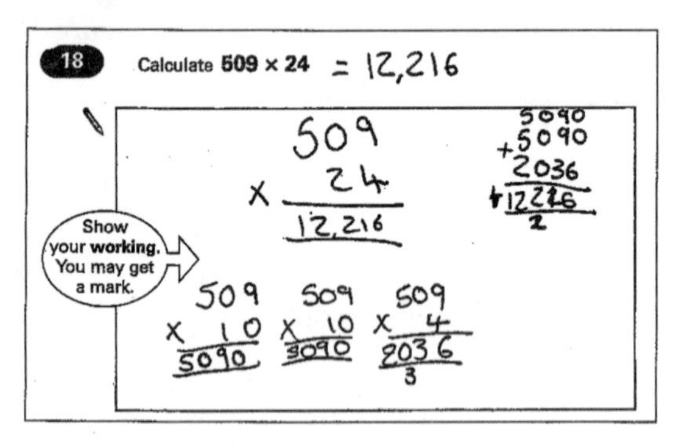

Figure 5.5 An effective method for calculating 509 × 24 (QCA, 2002)

Building methods for division

Inverse of multiplication

As the chapter on understanding the operations makes clear, division and multiplication are related as inverse operations and the result of a division calculation can always be related to a multiplication fact. It is expected that children will solve calculations such as 24 ÷ 8, 56 ÷ 7, 420 ÷ 6 not by doing a division procedure (either sharing or grouping) but by using associated multiplication facts. This will provide the most effective strategy when children have memorized the benchmark multiplication facts and should be the objective in teaching division calculations of this type. Understanding the connection between division and multiplication can help in solving missing number questions for example, □ ÷ 4 = 5 where the answer is found in 4 × 5 = 20. It may be easy to 'spot' the solution to this question but the similar problem '□ ÷ 21.7 = 37.5' from the 2004 national test paper could not be solved intuitively as there was no obvious connection between the numbers. As a calculator was allowed for this question an effective approach would be to use the connection among the triple of numbers related to multiplicative thinking so the answer to □ ÷ 21.7 = 37.5 could be found by multiplying 21.7 × 37.5 using the relationship between multiplication and division discussed in Chapter 3.

With whole numbers where the number to be divided (the *dividend*) is not a precise multiple it is helpful to use the idea of close numbers using an empty number line to give a supporting visual image. In calculating 53 ÷ 7, for example, the number 53 can be *located* between 49 and 56 which are the closest multiples of 7 (Figure 5.6).

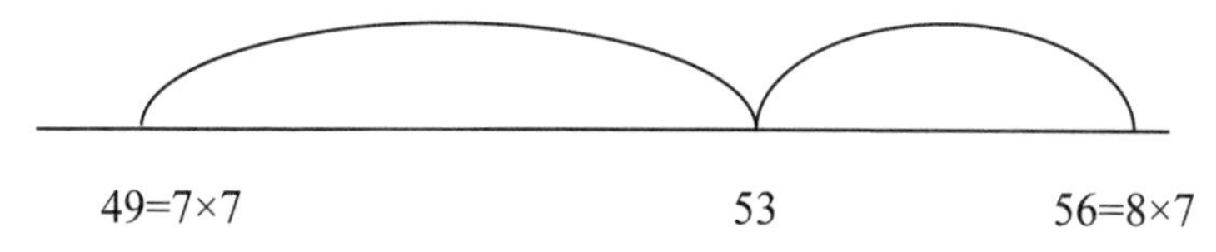

Figure 5.6 Identifying multiples of 7 that are close to 53 on an empty number line

Dealing with the difference between these multiples of 7 and the number 53 will depend on the purpose of the calculation or the context to which it relates. If '53 children are grouped into teams of 7' there will be 7 groups if 4 children have no group, or 8 groups where one group is smaller. The image of the empty number line can be helpful in introducing the idea that the remainder is $\frac{4}{7}$ and the result of dividing 53 by 7 is 7 and $\frac{4}{7}$ or just over 7 and a half.

Although the Framework (DfEE, 1999) is written with very few illustrations in context, the renewed Primary Framework (DfES, 2006a) makes it clear that remainders need to be dealt with according to the situation of the problem.

Guidance states that pupils should: 'Use practical and informal written methods to multiply and divide two-digit numbers (e.g. 13 × 3, 50 ÷ 4); round remainders up or down, depending on the context'.

(DfES 2006a: 12).

Using known facts

Chunking and partitioning for division

Some methods for division involve building up multiples of the *divisor* (the dividing number) while others are more closely associated with breaking down the dividend (number to be divided). It is here that number sense can make an important difference and experiences with breaking numbers into different chunks can provide effective mental methods. Although partitions according to place value will often be used for multiplication, for example 14 × 6 as 10 × 6 and 4 × 6, this will not always be most appropriate for division where alternative chunking may be more helpful. In order to do 98 ÷ 7 for example, it would be difficult mentally if it was thought of as 90 ÷ 7 and 8 ÷ 7, while looking for connections with the number 7 and chunking 98 as 70 + 28 and doing the calculations 70 ÷ 7 and 28 ÷ 7 would lead to a quicker solution. Change the calculation to 96 ÷ 6 and the chunks could be 96 = 60 + 36. Choice of good chunks will lead to greater efficiency in calculating and what children will come to notice is the way multiples of ten times the divisor can help.

Doubling and halving

Another way of making connections and simplifying a calculation may involve doubling or halving. When doubling is used in a division calculation it has different effects when it is used on the first number (the dividend) or the second number (the divisor). Look at the answer to the calculation 24 ÷ 4 and how it is related to 48 ÷ 4 and to 24 ÷ 8. When the first number is doubled then the total (the *quotient*) is doubled. When the second number is doubled, the quotient is halved. This is an important difference between multiplication and division. When these calculations are put into words: 'How many 4s in 24?' and 'How many 4s in 48?' or when images are used to show the relationships, it will be clearer to children how the result of one calculation may be used to find another. Using a context will also help exemplify what is happening: 'If apples are packed in boxes of 4 how many boxes will be needed for 24 apples?', then 'What if there were 48 apples?' or 'What if they were packed in boxes of 8?' When each of these calculations is recorded as a division the connections can become the topic for discussion and this is an example of conceptual discourse to support children's thinking.

Presented with other examples such as 68 ÷ 17 that are not immediately accessible it would be a very good strategy to first consider 34 ÷ 17, similarly 70 ÷ 5 could be related to the easier calculation 70 ÷ 10. These connections can be explored through web diagrams that children can construct and extend as an individual or a collaborative activity (Figure 5.1). Displayed on a classroom wall these can stimulate thinking and raise awareness of ways of working.

Sharing procedures

Although the relationship between multiplication and division will often lead to the most efficient calculating, some children will continue to associate division with the activities of sharing and grouping. Both of these ideas can lead to calculation procedures which reflect the understanding an individual has. Where they are the basis for understanding, working with the children's intuitive ideas and developing their efficiency can be more effective than trying too quickly to replace them by alternative efficient methods.

The illustration (Figure 5.7) shows a Year 5 (age 11 years) pupil's approach to a context problem that is worded so that it is likely to trigger a sharing method. The question for a teacher is how to interact with this pupil in order to move the calculation strategy forward without over-riding the understanding that it is based on. In this particular example it is clear that the pupil is moving in the right direction, having first shared 5 at a time, the next efforts are to share 10, then 15 and then 20 at a time. There appears to be an awareness that this still falls a long way short of the 1256 apples to be shared as the calculation is not completed. Intervening by asking 'Could the apples be shared 100 at a time?' could have been a *prompt* that raised a new possibility and led to improved efficiency without loss of understanding. It is also clear in this example that the written record the child has made is not helpful in supporting a sustained effort and this child would benefit from some discussion about ways to record calculations. The systematic

1256 apples are divided among 6 shopkeepers.
How many apples will every shopkeeper get?
How many apples will be left?

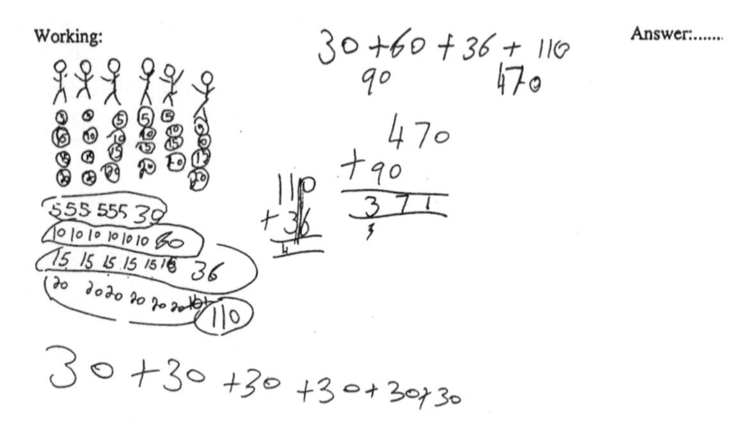

Figure 5.7 A sharing strategy for dividing 1256 by 6

432 children have to be transported by 15-seater buses.
How many buses will be needed?

Working:

Answer: 35 with

Figure 5.8 A grouping strategy for dividing 432 by 15

repeated subtraction of the chunking algorithm discussed in the next section could be an appropriate arrangement for this example.

Another example (Figure 5.8) based on a *grouping* context shows the efforts of a child who is clearly confident in understanding the problem and doing appropriate calculations.

The method is based on *building up* multiples of 15 to get as close as possible to 432. This is clearly a competent child with good understanding but the informal written record is messy and difficult to follow which may be the reason why it is not completed correctly. Once again it would be helpful to support this child in recording the calculation in some more structured way without losing the correct thinking that is involved.

Structuring a written record

Quite complex calculations can be done mentally or with the use of jottings. For complex calculations, however, children's own working can become messy to the extent that they lose their way and cannot complete the calculation. It is here that they need support in developing a structured record of their working. Research has shown that where Year 5 pupils were left working informally they were less successful than those who were supported in structuring their recording (Anghileri, 2006b).

Government guidance on the development from mental and informal methods to written methods makes it very clear that a progressive approach is needed:

> Over time children learn how to use models and images, such as empty number lines, to support their mental and informal written methods of calculation. As children's mental methods are strengthened and refined, so too are their informal written methods. These methods become more efficient and succinct and lead to efficient written methods that can be used more generally.
>
> (DfES, 2006c)

It is imperative that this progression to written methods nurtures understanding and puts value on an individual's informal thinking. Although very prescriptive in places this guidance makes it clear that learning procedures without understanding is no longer acceptable and it is stated that pupils need to 'develop and use written methods to *record, support and explain* multiplication and division … ' (author's emphasis).

The interpretation of a division such as 98 ÷ 7 as 'How many 7s in 98?' leads to an understanding of division as repeated subtraction. It would be possible to remove groups of 7 at a time, each time reducing the number left, but this would involve many subtractions and be very inefficient. Nevertheless it is worth illustrating this as a possible structured written method and asking children how they could make the process more efficient. Removing 14 at a time, that is 2 groups of 7, would result in better efficiency (Figure 5.9).

```
7 )98
   14  2×
   --
   84
   14  2×
   --
   70
   14  2×
   --
   56
   14  2×
   --
   42
   14  2×
   --
   28
   14  2×
   --
   14
   14  2×
   --
   00
```

Figure 5.9 Calculating 98 ÷ 7 by subtracting chunks of 14

Since the choice of chunks to be removed will be the child's choice it is worthwhile taking some time looking at different possibilities and making it clear that each individual can work at the level of efficiency they understand. The calculation above could be curtailed when 70 is reached if this is recognized as ten times 7. Alternatively it could be started with a chunk of 70 and then completed in less efficient or more efficient steps (Figure 5.10).

7)98		7)98		7)98	
14	2×	70	10×	70	10×
84		28		28	
14	2×	7	1×	28	4×
70		21		00	
70	10×	7	1×		
00		14			
		7	1×		
		7			
		7	1×		
		0			

Figure 5.10 Choosing bigger chunks for calculating 98 ÷ 7

The choice of chunks is an important skill that children need to learn. Since there are many different possibilities that will all work to give a correct solution, children can retain ownership of the calculation making their own decisions rather than trying to replicate the teacher's choices. In the traditional algorithm this is not the case and a wrong choice at any step will introduce errors (Anghileri and Beishuizen, 1998). This *chunking algorithm* is the written method proposed in government guidance on written methods for division which will be discussed later in this chapter.

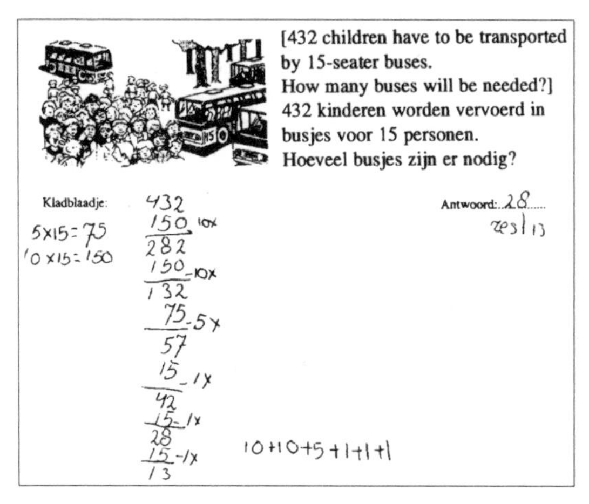

Figure 5.11 The chunking algorithm used to calculate 432 ÷ 15

As well as building progressively on the idea of repeated subtraction, and so reinforcing understanding, this chunking method will work even where the chunks are not the most efficient ones and the order in which they are used is not standard. It can introduce an opportunity for assessment as it reveals to the teacher the thinking of each individual and their progression towards efficiency. An example from a research study involving English and Dutch Year 5 pupils (Anghileri, 2001) shows clearly the choices made by this particular individual (Figure 5.11).

Following government guidance for written recording

DfES guidance on calculation (DfES, 2006c) proposes stages in developing written records for multiplication and division calculations. Building on the mental methods identified above, a structured record gives children a way of organizing their working but teachers need to be aware that it is unlikely that all children in the class will be ready for the same stage of recording. The guidance for multiplication culminates in the traditional algorithm but incorporates alternatives such as the grid method in acknowledgement that some children will prefer to use a written record they are confident with and understand. At stage 5, as well as the grid that relates to the area model for multiplication, there is a grid that is reorganized to relate more closely to the partition used in the al-

For stage 1 the method is mental multiplication using partitioning recorded horizontally:

$14 \times 3 = (10 + 4) \times 3$
$= (10 \times 3) + (4 \times 3) = 30 + 12 = 42$

$43 \times 6 = (40 + 3) \times 6$
$= (40 \times 6) + (3 \times 6) = 240 + 18 = 258$

or in a schematic form:

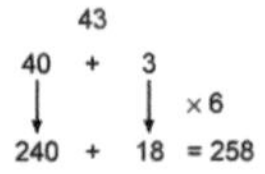

43
40 + 3
↓ ↓ ×6
240 + 18 = 258

At stage 2 the recording is organised in a grid:

$38 \times 7 = (30 \times 7) + (8 \times 7) = 210 + 56 = 266$

×	7
30	210
8	56
	266

Figure 5.12 DfES guidance on written methods for multiplication: stages 1 and 2

At the next stage 3 the grid lines disappear but the method is effectively the same and this is referred to as 'expanded short multiplication':

```
 30 + 8                        38
×    7                       ×  7
   210    30 × 7 = 210        210
    56     8 × 7 =  56         56
   266                        266
```

Now short multiplication is introduced as stage 4. Working with digits this gives a conceptually much more complex method with only a minimal gain in conciseness.

```
  38
×  7
 266
  5
```

Figure 5.13 DfES guidance on written methods for multiplication: stages 3 and 4

gorithm and consequently relates to more complex imagery. The extracts (Figures 5.12 to 5.20) are only part of the guidance documentation where additional explanations are given.

For stage 5 there are four alternatives given for two-digit by two-digit products and it is here that more conciseness can be gained (Figure 5.14). The complex structure of what is being multiplied by what may influence the choice of a less concise but more explicit way of recording. It is also worth noting that some children have difficulty holding in memory subtotals involved in some calculations and in recording the 'carried one' shown in the column methods.

Stage 5 shows various layouts for recording:

56 × 27 is approximately 60 × 30 = 1800.

×	20	7	
50	1000	350	1350
6	120	42	162
			1512
			1

	50	6	
×	20	7	
	1000	350	1350
	120	42	162
			1512
			1

56 × 27 is approximately 60 × 30 = 1800.

```
   56
×  27
 1000    50 × 20 = 1000
  120     6 × 20 =  120
  350     50 × 7 =  350
   42      6 × 7 =   42
 1512
 1
```

56 × 27 is approximately 60 × 30 = 1800.

```
   56
×  27
 1120    56 × 20
  392    56 ×  7
 1512
 1
```

Figure 5.14 DfES guidance for written multiplication methods: stage 5

For multiplication there is a stage 6 for three-digit by two digit products, illustrated with the same recording as stage 5, for example:

286	
× 29	
4000	200 × 20 = 4000
1600	80 × 20 = 1600
120	6 × 20 = 120
1800	200 × 9 = 1800
720	80 × 9 = 720
54	6 × 9 = 54
8294	
1	

Figure 5.15 DfES guidance for written multiplication methods: stage 6

In all of the levels illustrated above there is a danger that the method introduced by the teacher will become procedural and used by children because they think it is best to conform to the standard one that is taught. This presents difficulties for teachers who are trying to establish some standardization while at the same time trying to keep individuals confident in tackling problems through their own thinking. In the final example illustrated in the DfES guidance above 286 × 29 is shown using a standard format where an easier calculation could be made by transforming the calculation to 286 × 30 and subtracting 286.

DfES guidance on written methods for division

For division there are also five stages identified culminating in 'long division' of a three-digit number by a two-digit number but using the chunking method discussed earlier in this chapter rather than the traditional long division algorithm (Figures 5.16 to 5.20). Although 'short division' is illustrated at stage 2 and stage 4 it may be best avoided as it does not fit with the progression towards the chunking method shown in later stages. What is not made clear at this stage is the conceptual leap from dividing whole numbers like 60 and 21 by 3 to working with digits where 80 is considered as the digit 8 in the first step but the remaining 2 is converted to 20 in the next step.

The chunking method gives a single procedure that works in all cases and allows the element of choice that enables children to retain ownership of the method and do it in a thinking way.

At stage 1: mental division using partitioning is expected with informal recording in horizontal format:

$$
\begin{aligned}
87 \div 3 &= (60 + 27) \div 3 \\
&= (60 \div 3) + (27 \div 3) \\
&= 20 + 9 = 29
\end{aligned}
$$

and alternatives are given for more schematic recording:

84

70 + 14

↓ ↓ ÷7

10 + 2 = 12

or using a grid relating division to multiplication:

×		
7	70	14

→

×	10	2
7	70	14

10 + 2 = 12

Figure 5.16 DfES guidance for written methods for division: stage 1

For stage 2 short division of TU÷U is first recorded in horizontal format:

$$
\begin{aligned}
81 \div 3 &= (60 + 21) \div 3 \\
&= (60 \div 3) + (21 \div 3) \\
&= 20 + 7 \\
&= 27
\end{aligned}
$$

Then short division method is recorded like this:

$$
3\overline{)60 + 21}\quad \text{quotient: } 20 + 7
$$

This is then shortened to:

$$
3\overline{)8\,{}^{2}1}\quad \text{quotient: } 2\,7
$$

Figure 5.17 DfES guidance for written methods for division: stage 2

```
  6)196
  - 60   6×10
   136
  - 60   6×10
    76
  - 60   6×10
    16
  - 12   6× 2
     4     32
Answer:   32 R 4
```

This is the chunking algorithm with rather inefficient but easily identified chunks. The total 196 is reduced in stages by subtracting multiples of 6. In the same 'stage' the most efficient chunks are shown.

```
  6)196
 - 180   6×30
    16
  - 12   6× 2
     4     32
Answer:   32 R 4
```

Figure 5.18 DfES guidance for written methods for division: stage 3

Stage 4 is initially shown in horizontal recording:

$$
\begin{aligned}
291 \div 3 &= (270 + 21) \div 3 \\
&= (270 \div 3) + (21 \div 3) \\
&= 90 + 7 \\
&= 97
\end{aligned}
$$

The short division method is then recorded schematically and a little more concisely:

$$3\overline{)290+1} = 3\overset{90\,+\,7}{\overline{)270+21}}$$

This is then shortened to include 'little numbers' that need careful interpretation as the following method involves the use of digits rather than whole numbers. It is at this point that the method becomes conceptually much more difficult and many children cease to understand the procedure or think of it as their method:

$$3\overset{9\;7}{\overline{)2\;9\,{}^{2}1}}$$

Figure 5.19 DfES guidance for written methods for division: stage 4

Stage 5 is illustrated first with the chunking method and the answer recorded below the calculation:

$$
\begin{array}{rrl}
24 & \overline{)\,560} & \\
20 & -\underline{480} & 24 \times 20 \\
 & 80 & \\
3 & \underline{72} & 24 \times 3 \\
 & 8 &
\end{array}
$$

Answer: 23 R 8

Figure 5.20 DfES guidance for written methods for division: stage 5

Stage 3 is illustrated as an 'expanded' method for HTU ÷ U and is back to the chunking method recorded in a schematic way (Figure 5.18).

Since stage 5 (Figure 5.20) which is called 'long division' again uses whole numbers in a chunking method it is questionable whether the conciseness at the last step is necessary. For most children it will be more effective to record division always as the chunking method which will work in every example and be more meaningful.

At all stages from stage 3 onward the chunking format can provide an appropriate structure for recording that is consistent with children's understanding. It is not as concise as the short division format but many children will prefer to work with whole numbers rather than digits and the lack of conciseness is made up for in its closeness to informal thinking and mental strategies (Anghileri et al., 2002).

An illustration of just how demanding the algorithm can be is in the example (Figure 5.21) from a pupil who achieved level 5 in the 2006 national assessments for mathematics gaining 39 out of 40 possible marks on paper A.

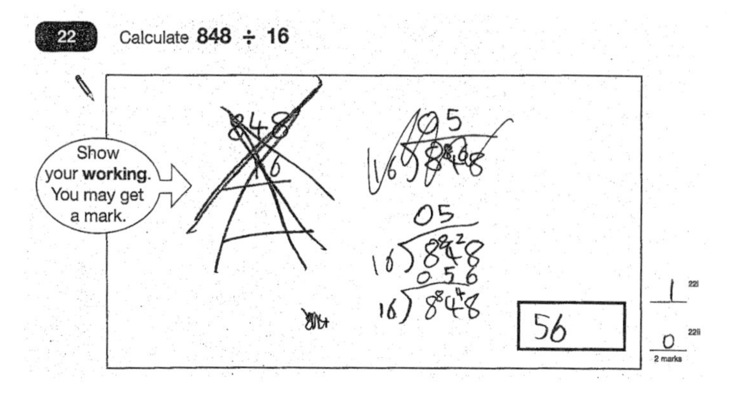

Figure 5.21 A nearly correct attempt at the division algorithm

A single mark was lost for a nearly correct division attempt. This able pupil appears to be dividing by 16 by dividing first by 8 and then halving. In the final step she appears to have calculated 6 eights in 48 and then forgotten to halve this 6 showing how complex the procedure can become and defeating even the most able pupils. Perhaps with the chunking algorithm she would have gained full marks.

Proportional reasoning

Although some time has been taken in this chapter to focus on written methods for calculating it is important to stress that this is not the most important aspect in teaching multiplication and division. What will have more lasting importance is an understanding of the operations and how they are related. As children progress through Key Stages 2 and 3 the connections become important as the idea of a ratio replaces division procedures and *proportional reasoning* is introduced which incorporates both multiplication and division. The following example from a national test in 2001 involves easy calculations but also the conceptually more complex situation of proportional reasoning:

> Six cakes cost one pound eighty. How much do ten cakes cost?

There are two approaches to this calculation:

- the unitary approach – finding the cost of one cake as the first calculation and then using this to find the cost of ten cakes
- the ratio approach – dividing the cost by 6 and multiplying by 10 to find $^{10}/_{6}$ of one pound eighty.

These relationships can be illustrated in a ratio table that gives children a model to work with (Figure 5.22).

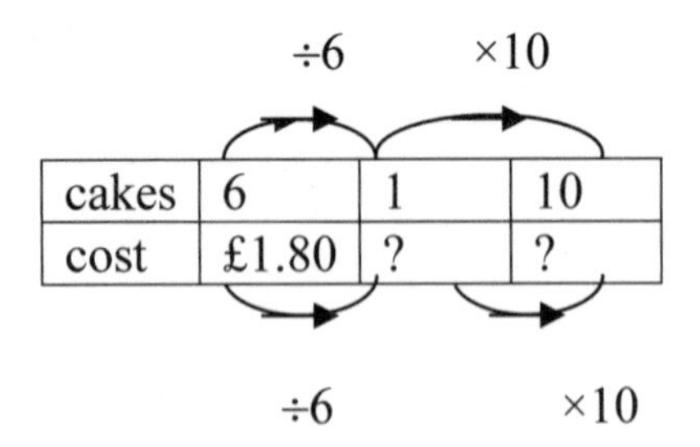

Figure 5.22 A ratio table for modelling a problem-solving strategy

In some examples a unitary approach will lead to more complex calculation as the example from Key Stage 2 national tests in 2004 illustrates:

> Mari is the presenter of a weekly radio show.
> She always plays **five** new songs for every **two** old songs.
> Last week she played 15 **new** songs.
> How many songs did she play **altogether**?

Clearly the calculations here are not difficult for Key Stage 2 children but understanding the idea of proportional reasoning is needed. Here the unitary approach is difficult to apply but the image presented in a ratio table provides a strategy for finding the number of old records played (Figure 5.23).

×3 (from 5 to 15)

new	5	15
old	2	?

Figure 5.23 A ratio table for solving the 'songs' problem

The table incorporates not only a ratio of 3 × between the columns but also a ratio of ⅖ between the rows so that in other examples a choice can be made for selecting the easiest calculation.

These examples illustrate that multiplicative thinking needs to be developed in a way that can be related to different situations. Too great an emphasis on written recording and not enough consideration of the meanings of multiplication and division will be unhelpful and may inhibit children's confidence to tackle novel problems with their own understanding.

CHAPTER 6

Calculating with Measures

All of the methods for calculating with numbers that have been discussed in previous chapters will be used where a problem is set in context. The introduction of units of measure will present additional considerations that will be identified in this chapter. There will be further discussion in the next chapter on decimals and fractions and in Chapter 8 which addresses other aspects of problem solving.

The curriculum has changed dramatically to meet the changing needs of society as described in Chapter 1. It has also been modified because units of measure in the UK have changed from imperial measures such as feet, yards, inches, bushels and hundredweights, that incorporated numbers in many different bases, to predominantly base ten systems. Money, length, weights and capacity all use measures based on tens, hundreds and thousands although some imperial measures, such as pints and miles, still persist. The Primary Framework states that by Year 6 most pupils should 'solve problems by measuring, estimating and calculating: measure and calculate using imperial units still in everyday use; know their approximate metric values' (DfES, 2006a: 96).

Many experiences with measures will be gained outside school but there are some children who will have little practical experience and the classroom will provide their main opportunities. Where activities are purposeful, such as managing a feeding schedule for a pet guinea pig, or measuring a plot and planning a garden, children's motivation to understand and work with appropriate degrees of accuracy will be enhanced. As this book is about number, in this chapter the practicalities of measuring will not be addressed but it will be assumed that the calculations involved will be based on real experiences inside and outside the classroom.

Money calculations

Before discussing the modern curriculum it may be interesting to see an excerpt from a report on arithmetic teaching in 1932 which identifies the type of calculation that was expected in the 'Junior School Course' and the way that older systems of units involved fiendish complexities. *The Teaching of Arithmetic and Elementary Mathematics* (Hemmings, 1932) reports that:

> the column method of arranging long division has much in its favour, and is here recommended partly because of its neat and compact form and also because it avoids putting down multiplication sums ... which should be done mentally.

The report then shows a money calculation (Figure 6.1).

Example - £538, 14s. 1½d. ÷ 52

	£10,	7s.	2¼d.	
52)	£538	14s.	1½d.	
	52	360	120	68
	18	374	121	70
		364	104	52
		10	17	18

Answer. £10, 7s. 2¼d. Remainder 18 farthings

Figure 6.1 A money calculation from 1932 (Hemmings, 1932: 61)

The curriculum today does not focus on such standard calculations but requires pupils to 'choose, use and combine any of the four number operations to solve word problems involving numbers in "real life", money or measures of length, mass, capacity or time, then perimeter and area' and 'choose and use an appropriate way to calculate and explain their methods and reasoning' (National Curriculum 2006 online). In many ways this is more challenging as the solution strategies to be used are not specified and assessment tests often involve problems of more than one step so that decision making about the way to tackle a problem becomes paramount.

The money calculation above shows clearly the convention of commas and full stops that was required in the old system of recording money. The modern decimal system also has conventions that children need to learn such as writing £52.50 and not £52.50p or £52.5. This can confuse children as illustrated in the following response to a 'calculator allowed' assessment question in 2006 which will have gained no marks (Figure 6.2).

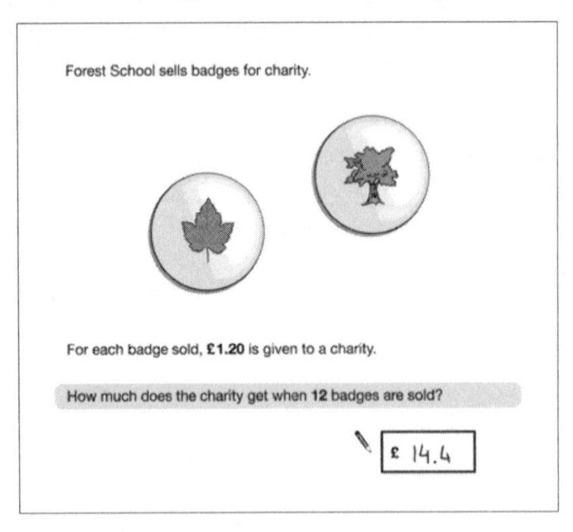

Figure 6.2 An inappropriate response to a money question (QCA, 2006a)

Although children live in a society that has money firmly established in everyday living, the extensive use of credit cards and electronic shopping can mean that many children do not have the same experiences with coins as previous generations. Classroom activities using real coins will support the familiarity that is needed to learn about values of coins and exchanges that are possible. Some recording will be necessary so that the symbol conventions can be learned and used appropriately. Collecting for a charity or running a sponsored event will provide a motivating task involving handling and recording different amounts, perhaps recorded on a spreadsheet. Puzzles and games such as the 'Money maze' on the nrich website (found by using the word 'money' to search the nrich site www.nrich.mathematics.org.uk) will provide investigative problems.

Calculations involving money may be identified with decimal calculations involving two places of decimals or may involve a conversion to a common unit so that the calculation can be completed with whole numbers. This type of decision making will be part of the planning strategy that an individual needs to develop for such problem solving. The division problem included in the Key Stage 2 (calculator allowed) paper in 2002 was about selling programmes for a concert:

> Programmes cost **15p** each.
> Selling programmes raises **£12.30**.
> How many **programmes** are sold?

With a calculator available (but not always used) it was necessary to identify the operation as division and to convert 15p to £0.15 or £12.30 to 1230p so that units were the same for the calculation. Mental calculation using multiplication could have been used since two programmes cost 30p and twenty programmes cost £3 giving quick access to the solution 82 programmes. This is an example where building up in chunks to a total using multiplication can be a more effective strategy than division. For calculations with money a partition method may also be favoured that treats the pounds and the pence separately with adjustment to the final total. In this case it is sensible to deal with the pounds first as this will give a good approximation that reflects best the overall value of the solution.

Assessment questions in national tests have tended to give information about the prices of particular items, sometimes from an illustrated list, and required children to select appropriate numbers and appropriate calculations. The calculation in Figure 6.3 was part of a question and a response reported in QCA 'implications for teaching' on problem solving (QCA, 2006a) suggesting that lower achieving children need to be encouraged 'to persist when solving problems of more than one step and to check that their answers are reasonable in the context of the problem'. The first part of the question asked

> How much **more** do the boots cost than the trainers?

and with a calculator allowed this was more about interpreting the question and using appropriate data than about the calculation itself.

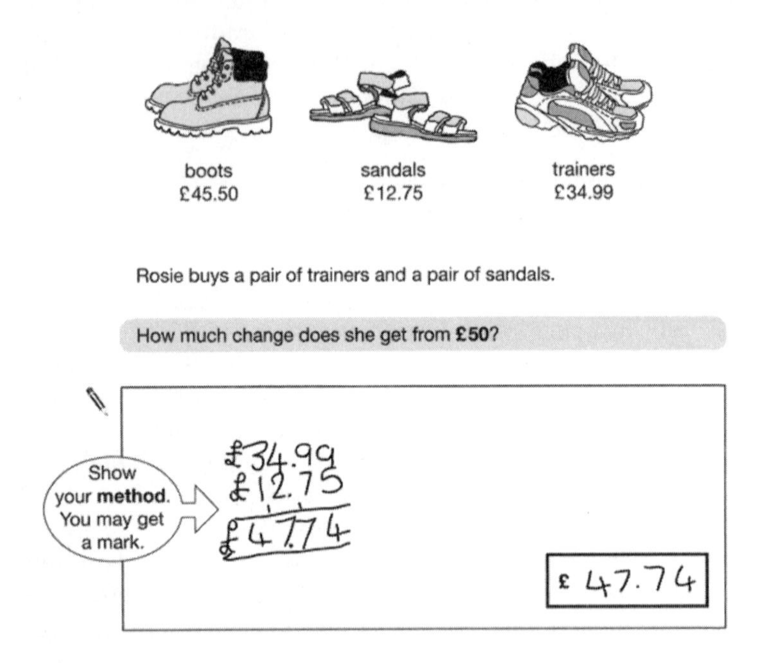

Figure 6.3 An incomplete response to a 2006 Key Stage 2 national test question

'The most common method used was an addition and subtraction method, although only just over 45 per cent of pupils working at the target level (level 3 for this particular question) recorded a complete method leading to an answer' (DfES, 2006c). This suggests lack of experience in tackling multi-part questions and may reflect a difficulty in knowing the 'right' way to record a solution. In tackling multi-part questions it becomes important to plan a strategy and to be aware that there will be no standard way to record any working. This lack of a specified written method for multi-part questions can inhibit some children as will be discussed in Chapter 8.

Multiplication and division problems in contexts involving money or measures will involve similar difficulties to those identified above for addition and subtraction. The conversion of units will be paramount as shown in the following Key Stage 2 national test question from 2003:

> Cream cheese costs £3.60 for 1kg.
> Robbie buys a pot of cream cheese for 90p.
> How many grams of cheese does he buy?

Having converted the two amounts of money to the same units, 360p and 90p, a common answer was 4, arrived at by correctly calculating 360 divided by 90 (or recognizing that 360 is found by doubling 90 twice) and failing to convert to the required unit of grams. This is another example of a multi-stage question involving money conversion and proportional reasoning where a ratio table may have helped (Figure 6.4).

£3.60	×4	1000 grams
90p		

Figure 6.4 A ratio table to support problem solving

In the assessment test the following year, 2004, a multiplication also involved money and a different difficulty arose:

> Cinema tickets cost **£3.65** each.
> Hannah buys **4 tickets**.
> How much does Hannah pay?

Since this was on the calculator allowed paper the calculation could be done easily but like the 2006 question above, the display showed £14.6 which was not always interpreted correctly as £14.60, which was necessary for the mark to be gained.

While children will have experiences with money in their daily lives this rarely involves writing down amounts or recording calculations. Much of the everyday working with money is done mentally and sometimes uses approximations. This is a good basis to begin school work with money which can extend children's understanding and focus on ways to record calculations that will be necessary in assessments. It is not only for the purpose of assessment but also a good experience in preparation for personal account keeping and planning as each individual becomes more financially independent.

Time calculations

Time is another measure that is evident in children's everyday lives but again this rarely involves any written calculations. It is a good example of a measuring system that is not based on units of ten and it is highly unlikely that we will ever change from 24 hours in a day and 52 weeks in a year. In addition to working with hours and minutes, time calculations will involve measures of weeks, months, calendar years and even centuries. Each of these measures can be identified with a time line which will be a particular example of a number line with specific calibrations. In the case of weeks and months, a calendar shows the regularity of weeks and the irregular pattern of days in each month. Planting seeds or monitoring weather patterns in science are examples of activities that will involve these measures. Associated problem solving is often done with a calendar available and as most adults will continue to use this support for calculating, one should be readily available in the classroom.

Calculations with time measured in years and centuries can provide contexts for working with four-digit numbers using events such as the children's own birth year, or the timing of significant events such as the year man landed on the moon. Linking to historic or scientific events will integrate different curriculum subjects with mathematics and use calculations in a meaningful way. Such associations can also provide the motivation for constructing a time line with four-digit numbers in their relative positions. Here all the experiences of locating numbers, identifying close numbers and benchmarks and constructing jumps between numbers, which have been discussed in earlier chapters, can be used.

In the case of hours and minutes a number line on the clock-face will come in the form of a calibrated circle with some or all of the numbers from 1 to 12 missing. Children will need a mental image of the 60 minutes that pass as the 'long hand' makes one revolution and where 25 minutes, for example, will be identified by the position of this long hand 5 minutes before the half hour. This circular number line on an analogue clock-face provides visual images for understanding 'quarter to' and 'half past' and 'quarter to' which are not available on a digital clock-face. On the other hand the time can be 'read' from a digital clock while it often has to be 'interpreted' from the incomplete scale of an analogue clock. Interpretation of a different kind is needed to make sense of digital time and this can provide some meaningful calculations around 60 and 30. Knowing that 10.57 is just 3 minutes before 11 o'clock or that 3.29 is just before half past 3 are both examples of real situations adding to children's sense of numbers.

In many problems that are set within a context the first demand is to extract the relevant information and identify the appropriate operations for combining the numbers. Calculations involving hours and minutes are often in the form of finding differences in time and the idea of time changing suggests a sequence method for calculating. The illustration (Figure 6.5) shows a number line being used to support such a calculation.

A successful strategy used by some children was to record and calculate the lengths of time taken for both Claire and Tim in order to work out the difference between the two times.

Here are the **start** and **finish** times of some children doing a sponsored walk.

	Start time	Finish time
Claire	9:30	10:55
Ruth	9:35	11:05
Dan	9:40	11:08
Tim	9:45	11:05

How much longer did Claire take than Tim?

5 **minutes**

Test B, Q17, example of a correct response.

Figure 6.5 An empty number line being used in a calculation with time

The jumps on the number line can be identified with intervals of time, for example 30 minutes or 5 minutes, which may help to reinforce the 60 minute interval between the hour calibrations in contrast with 10 or 100 calibrations more usually found. In national tests a common children's mistake is for the calculation to be based on intervals of 100 giving the wrong solution.

Problems with time can also be formulated to involve addition or multiplication. At Key Stage 2 there are problems similar to the following example from the 'calculator not allowed' paper in 2004:

> A film starts at 6:45pm. It lasts 2 hours and 35 minutes. What time will the film finish?

This sets addition in the context of time and will require 'chunking' in a different manner from abstract numbers as well as several steps in the solution. The use of two notation systems '6:45pm' and '2 hours and 35 minutes' requires adjustment so that 45 minutes and 35 minutes are identified and added to a total that is 1 hour and 20 minutes. Perhaps with a jotting to help remember this partial total, the number of hours is now calculated as 6 + 2 + 1. The resulting 9 hours and 20 minutes then needs to be given as a time in the format 9:20pm for a full solution to the problem.

Working with aspects of time that are pertinent to children's own lives, for example monitoring progress through the school day, or planning television viewing from published programmes, will provide opportunities to discuss ways of calculating and should incorporate some recording in appropriate formats. This will include the use of am and pm as well as the 24-hour clock. There is evidence from national tests that difficulties with these ideas may persist into secondary school as seen in responses to the mental mathematics question from the 2003 national assessment:

> It takes three hours to travel from my home to my friend's house. I arrive at 2pm. At what time did I leave home? Write your answer using am or pm.

Nearly a tenth of pupils gave an answer of 12am, midnight or equivalent which suggests that they counted back incorrectly and confused morning and afternoon. A similar proportion gave an answer of 11pm or 23:00. These findings suggest that many children need more experience of interpreting and using times recorded in many different ways including the use of am and pm.

Weight and capacity

The National Curriculum refers to 'mass' rather than 'weight' as this is the correct scientific way to refer to the property of an object that remains independent of gravity. In practice the activity of 'weighing' objects leads to a numerical description based on units of measure that are usually kilograms and grams. Although pounds and ounces can still be used in certain purchases the decimal weights are steadily replacing them and by the time the present generation of children leave school it is likely that such standard units will be little used. As indicated above, the Primary Framework (DfES, 2006a) requires children to be familiar with imperial measures still in daily use and know their approximate metric equivalents but they are assessed most often on their understanding and ability to calculate with metric measures.

Practical activities will give children a good understanding of the metric weights gram and kilogram and their vast difference makes it easier to remember that 1000 grams make 1 kilogram. This introduces complexities in recording weights as the same quantity can be recorded in different ways, for example 0.4 kilograms is the same as 400 grams. Understanding this conversion from one unit to another is central to many calculations with measures. The renewed Framework states that in Year 5 most pupils learn to 'convert larger to smaller units using decimals to one place (e.g. change 2.6kg to 2600g)' and in Year 6 most pupils 'select and use metric units of measure and convert between units using decimals to two places (e.g. change 2.75 litres to 2750ml, or vice versa)' (DfES, 2006a: 19). This ability to convert to different units will come from experiences recording practical activities and reading scales displaying more than one unit.

Many examples of different number lines are found in the calibrated scales used in weighing and measuring and these can be circular as well as linear. The aspect of interpreting information in a problem is relevant to reading the scales found on weighing machines. Because 1000 or more calibrations are not practical the intervals on different scales will vary and children's experiences with an empty number line will help them to locate numbers in relation to some benchmark numbers that are shown. The following response to an assessment question (Figure 6.6) shows some competence in reading a scale and in calculating but lack of understanding of the implications of the units that are specified (QCA, 2006b).

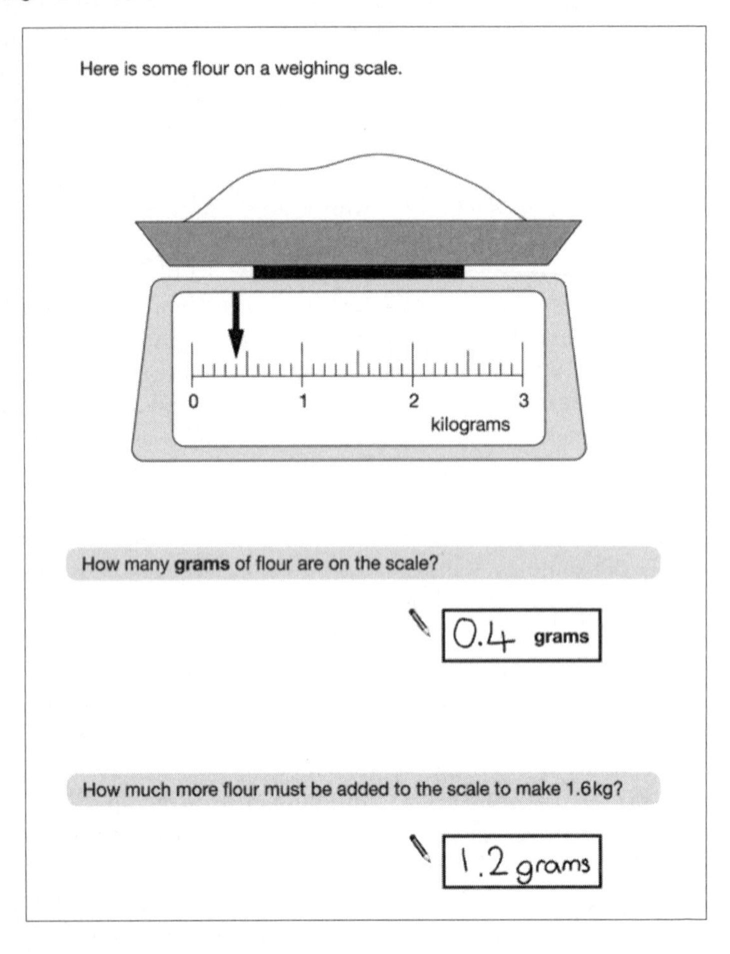

Figure 6.6 Reading a partially numbered calibrated scale

In this illustration the interval between each kilogram mark is divided into ten parts with the halfway mark between kilograms accentuated. On other scales the same interval between kilograms is commonly divided into four or five parts requiring children to be attentive to the differences and thoughtful in interpreting the data. Guidance also specifies that children need to be able to 'Interpret a reading that lies between two unnumbered divisions on a scale' (DfES, 2006a: 34).

Calculations with weights may be best done either with all the weights converted to grams, often with large numbers involved, or with kilograms that are represented using decimals. Part of the decision making children will engage in when solving problems with weights is how to convert the measures to units that are convenient to calculate with. Converting can cause problems as illustrated in the 2002 Key Stage 2 national test question which required children to

> convert 0.03 kilograms to grams

and where a common error was to give the answer 3g. More experiences with estimating the weights of objects, using scales with different calibrations, and even extracting and using data from packaging such as cereal packets and chocolate bars will help children get a 'feel' for these weights and their relationships.

There are several options for representation that include mixed units, for example the same weight can be represented as 1 kilogram and 200 grams, 1200 grams or as 1.2 kilograms, and these can be written as 1.2kg or 1200g. The decision about which representation to use will influence the calculation method to be used. In the first case there will be a 'natural' partition into the different units while in the case of 1.2kg the methods for calculating will be the same as those used for decimals. As with money, a calculator is a useful aid for extended or complex calculations and children will need to be able to enter data accurately in an appropriate format and to interpret the output, making jottings to keep track of the calculations.

Closely related to the units for weighing are the units of litres and millilitres for measuring capacity. As 1000 millilitres are equivalent to 1 litre the representations and calculations will introduce issues of using scales and representing quantities that are identical for those of weighing. The only difference may be the introduction of the unit decilitre for one-tenth of a litre which is more convenient for use in the classroom than a millilitre but which is much less commonly found in popular use. Although gallons and pints are still in common use they are found less frequently and no longer need to appear in calculations in the classroom.

Calculations with capacity again often involve conversions from one unit to another. The most usual units are litres and millilitres so that representations involve three decimal places. Where children have practical experiences of measuring liquids they will become familiar with benchmark conversions that 500ml is half a litre and 250ml is a quarter of a litre. Bottles of squash and fizzy drinks come in containers marked with these and similar amounts so that children can develop a real sense of these quantities. In the classroom it will be necessary to identify the alternative representations using decimals so that children gain experience in converting from one system of units to the other and recording data using appropriate representations. Half a litre is represented as 0.5 litres and this is the alternative for 500ml but does not give a good sense of the 500 thousandths it represents. It is helpful to introduce one-eighth of a litre as 125ml or 0.125 litres – a bit more than one-tenth of a litre – to see the relevance of the three places of decimals. A 2003 Key Stage 3 question on the calculator allowed paper:

> A glass holds **225ml**.
> An adult needs about **1.8 litres** of water each day to stay healthy.
> How many glasses is that?

Here again, as a calculator was available the question was assessing the skill of identifying the appropriate operation and converting the two quantities to the same units.

Length – including perimeter and area

Most closely identified with classroom measuring activities are rulers and other instruments for measuring lengths and distances. Children's experiences with a 30cm ruler and a metre rule will support their understanding of the number line and can be the basis for purposeful activities in representing numbers with one or two places of decimals. This practical work will help with their understanding of tenths and hundredths and can give them experience of conversions from centimetres to metres and vice versa. This ability to convert from one unit to another will be closely associated with children's experiences with numbers and their understanding of the number system as it is extended to decimal notation. Work with kilometres will introduce the idea of thousandths and that 1m = 0.001km and curriculum guidance for primary school states that pupils should 'convert between centimetres and millimetres or metres, then between millimetres and metres, and metres and kilometres, explaining methods and reasoning' . . . . and 'locate on a number line, and order, a set of numbers or measurements; then recognise thousandths (only in metric measurements)' (DfES, 2006a). Work with an empty number line, locating numbers and calculating using this imagery will support understanding of calibrated scales and practical measuring activities which will, in turn, support use of the empty number line.

As with money, there are issues relating to the different ways the same quantity can be represented but some implications for degrees of accuracy. Unlike money, both 1.4m and 1.40m are acceptable ways of recording, with the latter indicating a more accurate measurement correct to two places of decimals, or the nearest centimetre. For calculating it will again be necessary to identify the most appropriate representation to work with and will usually necessitate a conversion to make the solution meaningful in the context of the problem.

Calculations with length are sometimes associated with finding the perimeter, particularly of rectilinear shapes, using addition and multiplication or doubling. Once again there will be a choice of whether to work with whole numbers by converting each measurement to centimetres, or to work with two places of decimals and measurements in metres.

There is widespread confusion between perimeter and area and references are often together as in the National Curriculum:

> Pupils should be taught to: find perimeters of simple shapes; find areas of rectangles using the formula, understanding its connection to counting squares and how it extends this approach; calculate the perimeter and area of shapes composed of rectangles.
>
> (National Curriculum 2006 online)
> www.nc.uk.net

Once again, practical activities measuring areas using actual square centimetres and then counting squares on squared paper will help identify area as a different measure from perimeter. A square metre can be used in science activities such as sampling plants in a particular environment and a fun activity in the playground is to see how many children can stand on a square metre patch. These experiences provide imagery that can help interpreting measurements and considering appropriateness for solutions to later calculations. Progressing too quickly to calculating areas and perimeters at the same time will add to children's confusion. The same is true when volume is introduced and many children get little or no experience with practical measuring of volume. Handling cubic centimetres and a cubic metre, constructing or measuring objects with them, gives children a 'feel' for this quantity. Although inconvenient, it can be this very inconvenience that makes an activity memorable and useful. Another way to give children the 'feel' for the cubic centimetre is to identify the relationship between cubic centimetres, millilitres and grams. Metric measures are based on the relationship that one litre of water weighs one kilogram. One millilitre of water will occupy one cubic centimetre of space and will weigh one gram.

Other measures

Before leaving this chapter a brief mention will be given to other measurements that will be encountered in mathematics problems. Temperature presents opportunities for measuring on a calibrated scale including negative numbers. This number line can be used to show differences in temperatures and used to explain the way negative and positive numbers are added and subtracted. These are difficult concepts to use in the abstract as adding a negative number will be the same as subtracting a positive one: $7 + (-5) = 7 - (+5) = 2$ (see also the discussion about representing negative numbers in Chapter 2). This idea of addition rarely relates to temperatures but an alternative arises in receiving and giving (or losing) money. In this example the calculation would relate to £7 received and £5 given away with a net outcome of £2 remaining.

Subtracting a negative number can be related to the difference between two temperatures and will have the same effect as adding a positive one: $5 - (-7) = 5 + 7$. Seeing this as the difference between 5 and -7 on a number line will help with understanding while rules such as 'two negatives make a positive' will generally lead to confusion and lack of understanding. Multiplication and division of negative numbers introduce further complexities and discussion will be found in books on secondary mathematics as well as on many websites.

In this book on calculating it will not be appropriate to discuss angles at any length but the units of 360° in one complete turn gives another measuring system and the protractor usually used for measuring angles another form of calibrated scale. Identifying a quarter turn as 90° and 180° as a half turn will give more number associations that may be useful in some calculations.

CHAPTER 7

Decimals, Fractions and Percentages

When children are used to working with numbers by relating them to each other in ways that have been identified in earlier chapters, development of fractions, decimals and percentages will reflect familiar patterns of representing numbers and operating with them. The extension of the number system needs to be progressive and fractions and decimals can be incorporated from a very early stage when they are met in problem solving. Talking about halves and quarters will precede their symbolic representation and children will start to operate with them as real quantities and as images, such as half a cake, or the different shapes for quarters of a square that can be found in pattern making. Decimals are found in the way quantities of money are represented and will appear as children solve real problems or explore number relations with a calculator.

Additional skills have to be learned so that children can cope with these new systems of numbers and these include:

- reading and interpreting the symbols
- locating the numbers on a number line
- calculating with different representations
- solving problems involving various forms of numbers.

Without the appropriate connections it will seem to children that they have totally new systems of symbols and new rules to learn. Children need to learn how to move among the various possible representations in a flexible manner:

> if they (children) know that ½, 0.5 and 50% are all ways of representing the same part of a whole, then the calculations
>
> ½ × 40
> 40 × 0.5
> 50% of £40
>
> can be seen as different versions of the same calculation.
>
> (QCA, 1999: 52)

But each representation brings with it all the conventions for interpreting and operating with the symbols, and the idea that rational numbers (that is, any number that can be expressed as a ratio of two whole numbers) have equivalent representations as fractions, decimals and percentages is far from obvious. There are further numbers, the irrational numbers, which cannot be expressed exactly as a ratio but are the non-terminating, non-recurring decimals such as $\sqrt{2}$ and π.

It has been traditional to start teaching children about fractions and then extend to equivalent decimals and to percentages. Alternative approaches have been researched and Moss and Case (1999) report a successful teaching experiment beginning with percentages rather than fractions and decimals. Although this is by no means an exhaustive study it indicates possibilities that can be considered that break with tradition in an informed way. The justification for a change was to 'capitalise on children's pre-existing knowledge' of whole numbers and to postpone 'the problem of having to compare and manipulate ratios'. Because all percentages have an easy representation as a fraction, for example, 27% is $^{27}/_{100}$, conversion from percentages to fractions is straightforward and children can make conversions in a direct and intuitive fashion. The decimal equivalents of percentages are also straightforward as $^{27}/_{100} = 0.27$. This is not true of the conversion from fractions to percentages as simple fractions such as $^{1}/_{3}$ and $^{1}/_{7}$ do not have easy equivalents as percentages. Through this approach, starting with percentages, the children in the study developed 'better global understanding' and 'the sort of confidence, flexibility, and inventiveness . . . called for . . . in number sense' (Moss and Case, 1999: 143).

Traditionally fractions have been taught first because they were needed in measuring systems, particularly time, money and linear measure, with quarter and half hours, pounds, shillings and pence, feet, inches, yards and so on. In these systems it was often necessary to calculate with complex fractions as the money calculation in the last chapter shows. Now that so many measuring systems are based on units of ten they are better represented using decimals and decimal numbers are more often found in everyday use. As well as the benefits, the familiarity of everyday representation can bring with it problems of interpretation that will be discussed later in this chapter.

Starting with percentages

Justifications for starting with percentages include children's familiarity with them because of their preponderance in society today, and the fact that calculating with percentages will use the children's strategies for whole number computation. In the environment at home and in school percentages are found on packages and in advertising, and every computer has a '*percentage bar*' to show the stage of execution of some procedure. As children's everyday experiences provide contexts in which percentages appear they will often start with intuitive ideas of what different numerical values mean. Many appear to have understanding that 99% means 'almost everything', 50% means 'exactly half' and that 1% means 'very little'. Such intuitive meanings can help guide understanding of the equivalent forms of decimals and fractions, particularly when children work from familiar 'benchmarks' such as the equivalent forms for 50%, 25% and 10%. From these known percentages many others can be derived, for example 60% = 50% + 10%, 5% is half of 10%, 12.5% is half of 25% and 150% is 100% + 50% so that new percentages can quickly be found using addition and doubling or halving. This way of working from known percentages can give effective mental methods for calculating.

Visual images for illustrating percentages include a hundred square with portions shaded in various ways to give the children a sense of 'how many parts out of 100' and this can be used effectively to relate percentages to fractions and decimals. Another useful image is the percentage bar referred to above showing 0% to 100%. This can provide a familiar icon to help connect children's intuitions about proportions to the percentage scale. This percentage bar can be used to model the stages in a calculation, for example to find 17.5% of £24 by first finding 25%, 12.5% and 5% (Figure 7.1).

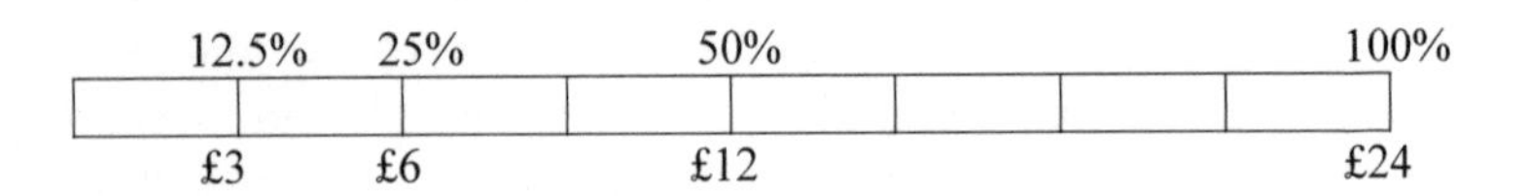

Figure 7.1 A percentage bar showing various percentages of £24

Research in the Netherlands advocates the introduction of realistic (meaningful) situations that explore the way percentages can be used and the depth of understanding that needs to be developed (van den Heuvel-Panhuizen et al., 1995). Their progressive development uses contexts such as a partially filled parking lot to introduce visual models of percentages that will be represented later as a percentage bar. Among the examples the Dutch have used to promote discussion of the way percentages are used is an example showing two pairs of shoes, one with the label **40% off** and the other with the label **25% off**. Discussion is encouraged to consider the different scenarios that could make the '25% off' pair a better bargain or a more manageable purchase depending on the original prices.

Establishing the meaning of decimal numbers

When children have a 'feel' for percentages and their relative size, this may be used to introduce decimals using the alternative representation with 10% as 0.1 and 1% as 0.01. These 'benchmarks' can provide meanings for the relative size of different decimals and can give children a 'feel' for such numbers. On the percentage bar 10% will be one tenth of the bar and 1% will be one hundredth of the bar (Figure 7.2).

10%	10%	10%	10%	10%	10%	10%	10%	10%	10%
0.1	0.1	0.1	0.1	0.1	0.1	0.1	0.1	0.1	0.1
$\frac{1}{10}$	$\frac{1}{10}$	$\frac{1}{10}$	$\frac{1}{10}$	$\frac{1}{10}$	$\frac{1}{10}$	$\frac{1}{10}$	$\frac{1}{10}$	$\frac{1}{10}$	$\frac{1}{10}$

Figure 7.2 A percentage bar showing equivalent percentages, decimals and fractions

This model has an advantage over the imagery of a hundred square because it will convert more readily to a number line with decimal numbers in their relative positions. For numbers less than one, decimal representation involves an extension of the place value system to include tenths, hundredths and thousandths, and without imagery to support their introduction, it can be difficult for children to interpret the symbols and to locate them in relation to each other. Another format of the percentage bar showing decimals between 0 and 1 will emphasize their ordering while shaded areas on a 100 square can show their relative size.

The convention of giving large numbers names relating to their size, for example 54 is 'fifty four' and 540 is 'five hundred and forty', is not extended to decimals. The number 0.54 is rarely read as 5 tenths and 4 hundredths and is usually read as 'nought point five four' involving language that does not associate the digits with their relative place value. There is a great temptation for children to 'read' the symbols as whole numbers, for example 0.54 as 'nought point fifty four' and consequently identify this as a larger number than 0.7. When asked if there is a difference between 4.9 and 4.90 research has shown that many children respond like Frances (age 11) by saying 'Yes, 4.90 is more' (Brown, 1982). Complexities also arise because two-digit decimals are commonly used for money and measures with 1.25 read as 'one twenty five' where the five may also be considered as a 'unit'.

Ordering decimals on an empty number line will help with imagery and the 'washing line' activity referred to in Chapter 2 can involve the whole class in locating different numbers. Working initially with numbered cards with one place of decimals, for example 0.7, 2.3, 1.5, 0.1 and 0.9, these can be positioned in relation to each other on a 'washing line'. The distances between any two of the numbers can be discussed and this marks the beginning of calculating with decimals. When numbers with two places of decimals are introduced some initial experiences with a number line calibrated in one-hundredths will be helpful. Such a number line will be found on a container marked in millilitres or on a metre ruler showing centimetres. Both the contexts of length and capacity can involve practical tasks that provide meaningful experiences in using two places of decimals. Recording results on a number line can then involve locating numbers with one and two places of decimals and this can lead to successful calculating as the following example from the Key Stage 2 assessments in 2002 shows (Figure 7.3).

This illustrates a successful solution using 'a self-drawn number line with the steps clearly marked' and an unsuccessful attempt using column arithmetic with 'an inappropriate strategy of taking the smaller digit from the larger' (QCA, 2003a).

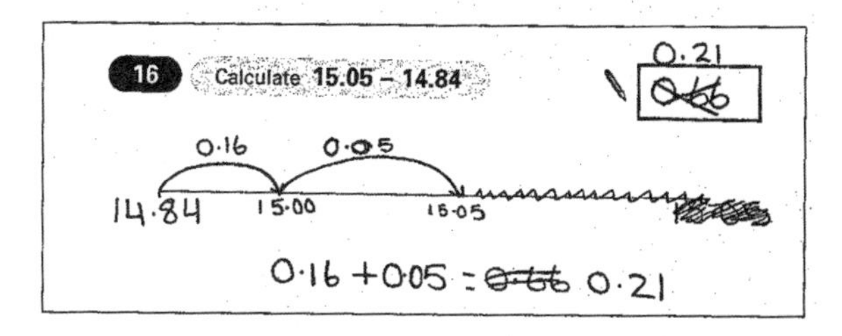

A common incorrect answer given by children working at level 3 was 1.81. These children used an inappropriate strategy of taking the smaller digit from the larger digit in any column.

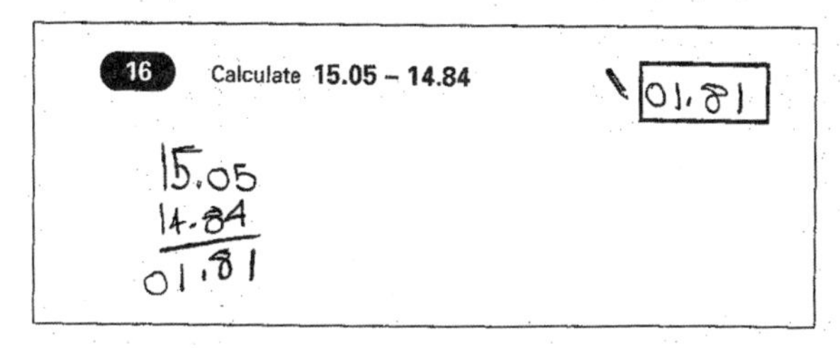

Figure 7.3 Calculating with two places of decimals on an empty number line

Multiplying and dividing by ten

Initially the decimal point is introduced to separate the whole part of a number from the fractional part. The idea that multiplying by ten 'shifts' all numbers to the left and dividing by ten 'shifts' all numbers to the right may be explained in terms of column headings (units, tens, hundreds ... extending to the left, and tenths, hundredths ... extending to the right) where the decimal point separates whole numbers from parts of a whole. This idea can cause difficulties not only because of the similarity of the column labels and the complexity of the word constructions, but also because the labels are meaningless until children have some appreciation of the meaning of tenths and hundredths. This is where associating one-tenth with 10% and one-hundredth with 1% can help children make sense of the relative size of decimals.

Since numbers rarely come with the column labels already in evidence, it is better to interpret decimal numbers by considering the position of the decimal point when looking at the effect of multiplying and dividing by ten. The idea of multiplying and dividing by ten may be associated with 'shifting the decimal point' (or more correctly shifting the numbers in relation to the decimal point) and activities involving multiplying and dividing by ten repeatedly using a calculator will help to establish the pattern of behaviour. This fits with the 'feedback loop' described by Rousham (2004) that involves children in reflecting on the output and learning through the patterns observed. The ability to see numbers such as 0.3 as one-tenth of 3 will enable calculations like 12×0.3 to be associated with one-tenth of 12×3. Alternatively some children may prefer to consider this particular calculation using partitioning as 10×0.3 plus 2×0.3 which they may understand more readily.

Flexibility is needed in reading decimals such as 1.25 because the .25 represents 2 tenths and 5 hundredths but at the same time this is 25 hundredths. If children are able to recognize such whole numbers as 1200 flexibly as 12 hundred as well as 1 thousand and 2 hundreds, these ideas can be related to the system of decimal numbers. Whole numbers such as 47 can also be written as 47.0 or 47.00 so that multiplication by ten can again be interpreted by shifting the digits to the left. If children do not understand the role of zero in relation to the decimal point difficulties can persist into secondary schooling. In a classic study, Brown (1982) found that asking 13-year-old pupils to 'multiply 5.13 by ten' resulted in the responses such as 5.130, obtained by 'adding a zero', or 50.130 which dealt with the two parts of the number (before and after the decimal point) separately.

Recurring decimals

In practical situations with measures, recording may involve one, two or three places of decimals as tenths, hundredths and thousandths are involved, particularly in the units of capacity (1 millilitre = 0.001 litre) and length (1cm = 0.01m and 1m = 0.001km). When using a calculator children will discover that some calculations entered on it will fill the display with many places of decimals and may exhibit repeating patterns in the decimal digits. Many of these fractions, for example one-third, have recurring decimal representations they may be familiar with, so that $100 \div 3 =$

33.333333 or $33\frac{1}{3}$. Others will not be recognized and some like $\frac{1}{7}$ = 0.1428571 will not easily display its repeating nature (to see that this decimal recurs look at $(1000000 \times \frac{1}{7}) - 142857$ on a scientific calculator or the calculator accessory on a computer and the next part of the decimal representation will be displayed). Rounding errors can present interesting outcomes, for example many calculators will calculate '1' '÷' '3' '×' '3' '=' as 0.9999999 instead of 1 but the same as three lots of 0.3333333 or 3 thirds. The calculator offers children the opportunity to explore some decimal number patterns although a spreadsheet has the advantage of displaying many numbers at the same time and providing a printout of the numbers involved. A spreadsheet can be used on an interactive whiteboard so that teacher and pupils together can investigate number patterns. An interesting activity for children to try is to generate on a spreadsheet the results of dividing 12 by 1, then 2, 3, 4, 5, 6, 7, 8, 9, 10, 11 and 12 and then explain their findings.

Operations with decimals

In the same way that number sense for working with whole numbers involves some number relationships that provide 'benchmarks' from which other results may be derived, similar benchmarks need to be established for working with decimals. In particular 0.1 = 10% = $\frac{1}{10}$, 0.25 = 25% = $\frac{1}{4}$ and 0.3333 = $33\frac{1}{3}$% = $\frac{1}{3}$ and these can be used to work out many other decimal values by halving, doubling, multiplying and adding.

Identifying decimals that add to one can establish links with whole number computation, for example by joining the decimals in the top and bottom line that add to a whole number (Figure 7.4).

0.2	1.75	2.5	0.37	2.8	3.62	54.9	6.3
1.2	0.7	2.25	0.8	23.1	6.5	0.63	1.38

Figure 7.4 Matching decimal numbers that add to a whole number

This type of task will present opportunities for working with a mixture of one and two places of decimals.

It is difficult to replace intuitive ideas with algorithmic procedures for calculating but a vertical arrangement with the decimal points carefully lined up can provide a calculation method that relates to vertical addition of whole numbers. It was this aspect that was crucial for the 2006 national assessment question:

Calculate 52.85 + 143.6

Providing pupils have the understanding to make an estimation and check the appropriateness of the solution, this can be an efficient written method when a calculator is not available. Using an empty number line has the advantage that the size of the numbers has meaning in relation to their position and the jumps from one to the other. This makes it more easily understood by some children and more accurate.

When it comes to subtraction the column method has added complications even where the numbers are carefully aligned, as the exchange of tenths to hundredths and units to tenths can be conceptually more difficult. Many children will understand better a complete number method as illustrated with the empty number line used to support a calculation as illustrated in Figure 7.3.

Multiplication and division with decimals

Experiences with whole numbers can establish meanings for multiplication and division associated with 'repeated addition', and 'repeated subtraction' or 'sharing'. Research has shown that such notions are behind the misconceptions that 'multiplication makes bigger' and 'division makes smaller' (Fischbein et al., 1985). When it comes to calculations with decimals, children may struggle to attach a meaning to such calculations as 8 × 0.4 and 8 ÷ 0.4 and when deciding on which operation to use for solving a problem, the wrong operation is often selected. When asked to 'Ring the one which gives the BIGGER answer: 8 × 0.4 or 8 ÷ 0.4', Brown (1982) found that 58 per cent of 13 year olds circled the first. It is here that associations with the alternative representations of 0.4 as $\frac{4}{10}$ can help to establish a more meaningful calculation, and flexibility in using different representations will add to the 'sense' that children make of such calculations. As with whole numbers, the key to many calculations can lie in the fact that multiplication and division are inverse operations. To calculate '8 ÷ 0.4' it is possible to ask what 0.4 must be multiplied by to give 8. In this case the answer comes from doubling and then multiplying by ten (or these operations in reverse) to give the answer 20 calculated mentally. Establishing inverse operations can be helped with calculator activities, for example the loop diagram (Figure 7.5). Children can show their understanding of the inverse operations when they complete an empty loop diagram or give a partially completed one to another child to complete.

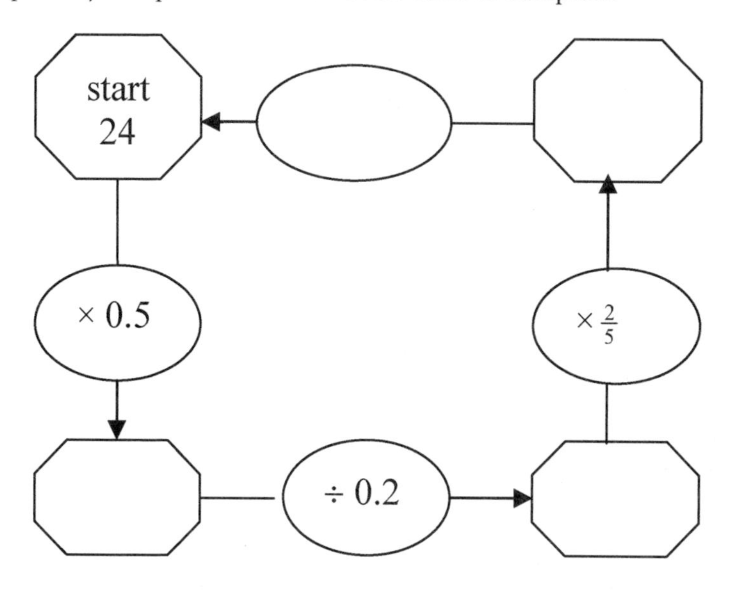

Figure 7.5 A loop diagram showing inverse operations with decimals

In a chapter entitled 'It's not calculators, but how they're used ... ' Ruth Forrester (2003) reports research on ways calculators can be used effectively and includes a variety of activities that work well in the classroom. One example is the following decimal game for two players. The first player sets a *target* number and the second player sets the *starting* number. The idea is to take turns to use the ' × ' button only to get from the start to the target. The winner is the first to get within 1 of the target number:

e.g. target 50, start 15
Game: 15 '$\times$ 3' = 45 '$\times$ 1.2' = 54 '$\times$ 0.9' = 48.6 '$\times$ 1.05' = 51.03 '$\times$ 0.99' = 50.52 WINNER!

(Ruth Forrester, 2003)

Chains of connections

Where decimal calculations are derived from calculations by halving and doubling, children will be better able to make sense of their meanings. 'Chains' of calculations can be recorded to show related facts (Figure 7.6).

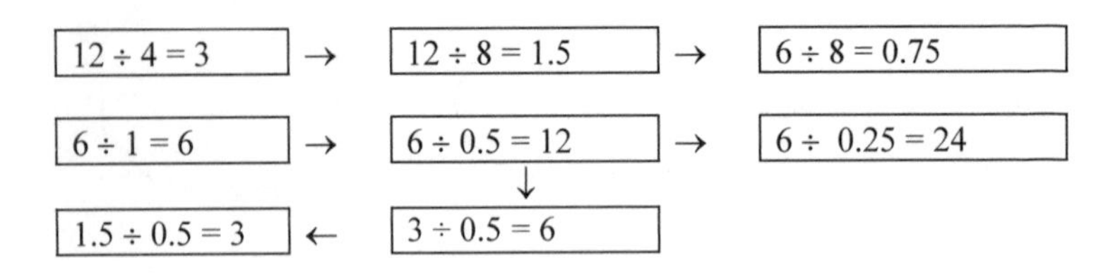

Figure 7.6 Chains of connected calculations

It can be helpful if children associate calculations such as 6×0.2 and $6 \div 0.2$ with images that will support calculating, such as:

- counting in 0.2s
- empty number line
- grid method
- calculator patterns
- links with fractions

In a similar way that 4×30 gives ten times the product of 4×3 so both 0.4×3 and 4×0.3 will give one-tenth of 4×3 and 0.4×0.3 will give one-hundredth of 4×3, that is 0.12. It will help children understand if these results are related to a visual image and a rectangular grid can be used to show which parts of a 10×10 square will result from 0.4×0.3. Jumps on a number line cannot easily be associated with this calculation but the grid method offers a meaningful extension (Figure 7.7).

When it comes to division of decimals $0.3 \div 0.2$ will be $\frac{1}{10}$ of $3 \div 0.2$ but will be ten times $0.3 \div 2$. As discussed in Chapter 5

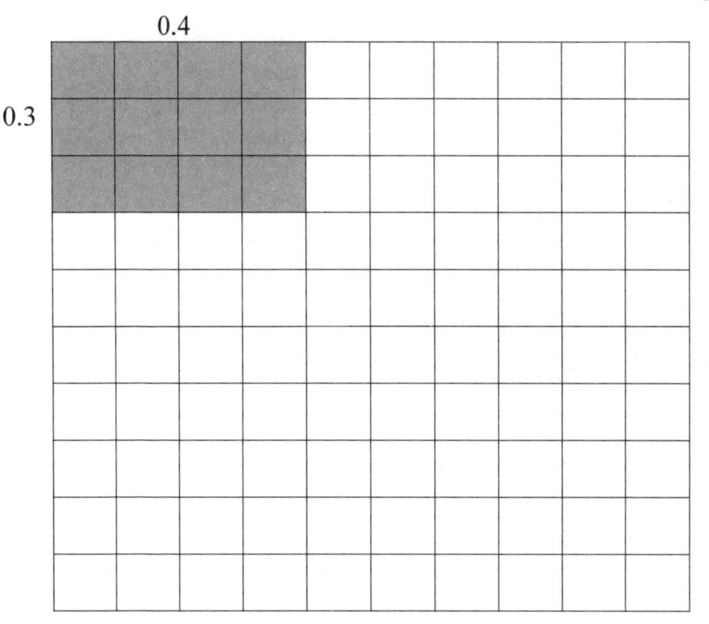

Figure 7.7 A grid showing 0.4×0.3 makes 12 hundredths

multiplying the first number by ten has the effect of multiplying the result by ten while multiplying the second number by ten has the effect of dividing the result by ten. In this way it can be seen that $0.3 \div 0.2$ is the same as $3 \div 2$. But a much better method is to relate this calculation to the fraction $^{0.3}/_{0.2}$ and find the equivalent fraction by multiplying top and bottom by 10. Using this strategy it will be possible to convert *all* division calculations involving decimals into equivalent calculations with whole numbers.

Introducing fractions

Much of the focus in school work is the identification of fractions as part of a whole or part of a collection. Early experiences with fractions involve practical activities including cutting and sharing with representation that may be in pictures and words before the symbols are introduced. Geometric arrangements of objects, for example pegs on a peg board, will then help children to relate fractions of a shape to a fraction of a collection and begin to use the language that will help them appreciate that the ratio between different numbers can give different representations of the same fraction, for example one half is the same as two quarters. As children come to understand the meaning of different fractions, oral work can include counting in fractions, for example, $\frac{1}{5}$, $\frac{2}{5}$, $\frac{3}{5}$, $\frac{4}{5}$, $\frac{5}{5}$, $\frac{6}{5}$ . . . with appropriate exchange for $\frac{5}{5}$ as 1 and $\frac{6}{5}$ as $1\frac{1}{5}$. This counting gives a verbal image that can be accompanied by images of shapes divided into portions, or by jumps on a number line. This counting will be more complex for counting in quarters or eighths where there are many more exchanges to be made but where students' understanding can be assessed by a listening teacher. In Chapter 1 there is a reference to a Key Stage 3 national assessment question where students were asked 'How many sixths in $3\frac{1}{3}$?' Only a third of pupils were successful although a strategy could have been to count up in sixths.

It is important to develop flexibility in using words and symbols and to identify links with the multiplication facts that children already know. Success with fractions will depend on the appreciation that a fraction represents both a number, and a ratio that reflects the procedure for finding this number. The symbol $\frac{3}{4}$ represents the number half way between $\frac{1}{2}$ and 1, three parts out of four from a whole, and also the result of dividing three by four. It is these latter identifications that enable children to understand that $\frac{6}{8}$, $\frac{15}{20}$ and $\frac{75}{100}$ all represent the *same number*. The latter representation will be crucial for identifying $\frac{3}{4}$ with 0.75 and with 75%. An approach to fractions which identifies each as numbers to be located on a number line, without emphasizing the way of partitioning a whole, will help establish the equivalence with decimals and percentages.

Locating fractions on the number line

Working initially with '1' on top (the *numerator*) the positions of fractions on the number line will involve a reversal of the children's normal understanding of relative size and will give a pattern where the distances between fractions (for example, between $\frac{1}{3}$ and $\frac{1}{4}$) are not unit distances as the children might expect (see Chapter 2). By keeping the number on the bottom (the *denominator*) fixed and varying the number on top (the numerator), children can learn to position various fractions, for example $\frac{1}{4}$, $\frac{2}{4}$, $\frac{3}{4}$, $\frac{4}{4}$, $\frac{5}{4}$ . . . and relate their ordering to the counting sequence for whole numbers. Working initially with numbers between 0 and 2, numbers will be introduced as 'top heavy' fractions and as 'mixed numbers' such as $\frac{6}{4}$ being the same as $1\frac{1}{2}$. By representing different 'families' of fractions on the same number line ideas of equivalence can be illustrated. Here again familiarity with multiplication facts is most crucial so that a fraction can be recognized in many different formats. Extension along the number line will provide opportunities to relate the different representations to multiplication facts where the ratios of numbers can appear in different forms, for example $\frac{3}{4}$ is the same as $\frac{9}{12}$, and as $\frac{15}{20}$ or $\frac{60}{80}$. As noted earlier in this chapter, some fractions will be easy to link with decimals and percentage representations, for example $\frac{1}{2}$, $\frac{1}{4}$, $\frac{3}{4}$, $\frac{1}{5}$ or $\frac{1}{10}$ and these will provide the 'benchmarks' from which other fraction equivalences can be derived.

Without being aware of it, limitations are imposed on mathematical understanding by conventions that appear to be well established in the classroom and it will take awareness and effort to open children's minds to the wealth of ideas associated with fraction representation. Research has shown that children have problems when asked to identify a fraction between $\frac{1}{2}$ and $\frac{2}{3}$ and in a study by Hart (1982) with 15 year olds only 21 per cent were able to give a correct answer. Mike Askew (2001) identifies one illustration of the way a social convention can restrict the possible mathematics within a situation (Figure 7.8).

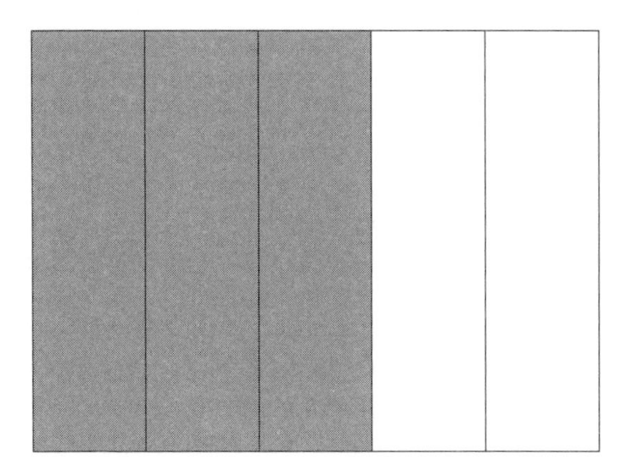

Figure 7.8 Fraction diagram

He suggests that if you ask teachers or primary school pupils what fraction is represented in Figure 7.8 most will say $\frac{3}{5}$ (or $\frac{2}{5}$ or possibly both). However, it is possible to 'read' the diagram in many other ways, including it representing $1\frac{2}{3}$, $2\frac{1}{2}$, $1\frac{1}{2}$ or $\frac{2}{3}$. He suggests that 'the reason it is almost universally read as $\frac{3}{5}$ is not to do with the diagram *per se*, nor to do with pupils' ability to perceive the fraction within the diagram'. Three fifths is taken as the common reading because this is a well established common practice: everyone from text book writers to teachers to parents 'reads' the diagram as three-fifths. Askew claims that it is a social practice at the heart of reading the diagram (Askew, 2001).

Calculations with fractions

For many generations calculations with fractions were practised by following learned rules, often with no understanding of why these rules work. To divide fractions, for example, the procedure 'turn the second one upside down and multiply' is still a mysterious process that cannot be explained by many adults. For children today, understanding is more important as it gives them confidence to think about the problems and to learn the relationships that will extend beyond arithmetic calculations and provide the foundations for algebra. Research has shown that formal procedures for adding and subtracting fractions have presented difficulties throughout schooling and Hart found that when 12–15 year olds were asked to add $\frac{1}{3}$ and $\frac{1}{4}$ a common error (made by 1 in 5 of the children) was to respond $\frac{2}{7}$, the result of adding the tops and adding the bottoms. She notes that this error occurred more when the problem was presented in computational form (as a sum) than when it was presented in a word problem (Hart, 1982).

Adding and subtracting fractions is only possible if they belong to the same 'family' with identical denominators (that is the numbers on the bottom are the same). To add $\frac{1}{2}$ and $\frac{2}{5}$, for example, alternative representations need to be found and this is where experiences with equivalent fractions are important. Even then, to add $\frac{5}{10}$ and $\frac{4}{10}$ without applying a 'rule' will only be possible if these numbers are recognized as '5 tenths' and '4 tenths', together giving '9 tenths'. It is better to discuss the meanings of these representations rather than introduce rules, such as 'add the tops', because children who understand will remember better the procedures and be able to transfer similar working to other situations. As the final step in the procedure, the resulting fraction, in this case $\frac{9}{10}$, must be adjusted, if necessary, to its *lowest* form with no common factors in the numerator (top) and denominator (bottom). It must be appreciated that children's knowledge of equivalent fractions will depend on their knowledge of multiplication facts and their ability to interpret flexibly the symbols they are working with.

Steps in the procedure for subtracting $\frac{1}{5}$ from $\frac{3}{4}$:

1. Find a 'family' that involves both fifths and quarters. This can be achieved by looking for a multiple of 5 that is also a multiple of 4.
2. Having found 20 as the smallest common multiple, change $\frac{1}{5}$ and $\frac{3}{4}$ to equivalent fractions in the family of 'twentieths'. By multiplying top and bottom of $\frac{1}{5}$ by 4, the ratio is preserved and the equivalent fraction $\frac{4}{20}$ is obtained. Similarly, multiplying the top and bottom of $\frac{3}{4}$ by 5 gives the equivalent fraction $\frac{15}{20}$.
3. Subtract '4 twentieths' from '15 twentieths' leaving '11 twentieths'.
4. Check that 11 and 20 have no common factors so that $\frac{11}{20}$ is in its lowest form.

If the resulting fraction was not in its lowest terms, for example $\frac{12}{20}$, an equivalent fraction is found by dividing top number and the bottom number by the same factors, in this case 4, reducing the fraction to $\frac{3}{5}$.

Multiplying and dividing fractions

The conversion to fractions in the same family is not necessary for multiplication and division but, starting with multiplication by a whole number, there are many stages in coming to understand the processes involved. At each stage it will be necessary to 'read' the symbols flexibly and discussion can be more helpful than working only with the written symbols on a page. Using the commutative rule for multiplication means that calculations such as $\frac{3}{4} \times 6$ and $6 \times \frac{3}{4}$ may both be interpreted as '6 lots of $\frac{3}{4}$' or 'three quarters of 6' and pictorial representation may be helpful in making this a meaningful calculation. Once again, interpreting $\frac{3}{4}$ as '3 quarters' will also be helpful in calculating the total of '18 quarters', and this may be identified with '4 wholes and 2 quarters'. Interpreting $\frac{3}{4} \times 6$ as 'three-quarters of 6' is not easy to reconcile with the symbol '$\times$' unless the phrase 'lots of' is adjusted to '$\frac{3}{4}$ of' and this is not immediately clear. The representations of '$\frac{3}{4}$ of 6' and '6 lots of $\frac{3}{4}$' will determine the same result. This result will also be obtained by 'multiplying 6 by 3 and dividing by 4' which is an alternative way of 'multiplying by $\frac{3}{4}$'. All of these interpretations give different ways to calculate the result and may be related to different illustrations.

The purpose in manipulating fractions in this way is not so much for usefulness in problem solving but more to explore the underlying relationships that exist within such calculations. When pupils meet algebra in secondary schools, all of these different interpretations will be identified with ways for manipulating abstract expressions, but in primary school the actual quantities and their images are helpful in identifying the patterns of behaviour as fractions are multiplied.

The next stage in multiplication involves fractions for both numbers and a meaning must be sought for interpreting calculations such as '$\frac{2}{3} \times \frac{1}{5}$'. The equivalent operations 'multiplying by one-fifth', 'finding one-fifth of' and 'dividing by 5' all lead to the result '$\frac{2}{15}$' which can be illustrated diagrammatically. It is also possible to apply the 'rectangle method' for multiplying starting with the representation of $\frac{2}{3}$ and $\frac{1}{5}$ of a shape. Once again the resultant $\frac{2}{15}$ will be the area shaded. The 'rule' that results is to 'multiply the tops' and to 'multiply the bottoms' and this may be established by considering the results of many examples involving different numbers. Working with the 'rule' and checking with an illustration will help to show that it will not always be necessary to consider the meaning and calculations can be undertaken with the confidence that they can be explained if necessary.

Dividing fractions

As with multiplication, division with fractions starts by giving meaning to division involving one of the numbers as a whole number. Since the commutative rule does not apply to division, the expressions '$4 \div \frac{1}{2}$' and '$\frac{1}{2} \div 4$' are quite different and will need to be interpreted separately. The expressions most easily understood will be different for each of these expressions with 'How many halves in 4?' and '½ divided into 4 (parts)' providing imagery that will enable each to be tackled by a different procedure. This ability to interpret a division expression in the most convenient way is at the heart of successful calculating where a whole number is involved. Some children will be uncomfortable with the idea that a division calculation can result in a number that is larger than either of the starting numbers as 'division makes smaller' can be an intuitive idea associated with division as sharing (Fischbein et al., 1985) or as repeated subtraction. For this reason, some time may be needed to give meanings to division expressions involving fractions so that the children themselves can be confident that they can find a meaning for each example. Using a calculator to look at patterns, for example dividing several numbers by ½ or ¼, can be helpful where the equivalent formats as the decimals 0.5 and 0.25 are understood.

In order to explain division involving two fractions, the idea of ratios can be used. In fact, where fractions to be divided, say $3\frac{1}{3} \div \frac{5}{6}$, are represented as a ratio $\frac{3\frac{1}{3}}{\frac{5}{6}}$ this can be converted to division by a whole number, in this case by multiplying the top and the bottom by 6. In this way any such calculation can be converted to division by a whole number.

For the more formal procedure that has been the basis for the taught procedure it is helpful to know about 'multiplicative inverses' as fraction numbers that 'undo' each other in multiplication. If the fraction $\frac{3}{4}$ is multiplied by $\frac{4}{3}$ the result will be $\frac{12}{12}$ or simply 1. The fractions $\frac{3}{4}$ and $\frac{4}{3}$ are called 'multiplicative inverses' as they multiply together to give the number 1. Similarly, any fraction can be 'undone' in this way using a fraction number with the top and bottom inverted, that is, the fraction 'turned upside down'.

Now consider the calculation $\frac{2}{3} \div \frac{3}{5}$.

Step 1: rewrite this expression as a ratio, $\frac{\frac{2}{3}}{\frac{3}{5}}$

Step 2: change this ratio into an equivalent fraction by multiplying the top and bottom by the same number. The number is chosen so that the bottom number becomes 1. That is, multiply the top and the bottom by $\frac{5}{3}$, the multiplicative inverse of $\frac{3}{5}$.

Step 3: now the calculation has become $\frac{\frac{2}{3} \times \frac{5}{3}}{\frac{3}{5} \times \frac{5}{3}}$ and this is equivalent to $\frac{2}{3} \times \frac{5}{3}$

This shows how division is achieved by 'turning the second fraction upside down and multiplying' but gives an explanation of why this works rather than offering it as a mysterious rule to be remembered.

Working with fractions or decimals

Since the systems of measurements today are predominantly metric, fractions appear less frequently in everyday situations than in the past and some children may prefer to work by changing fractions to equivalent decimals. This will be necessary if the calculation is to be undertaken with a calculator as few calculators work with numbers in fraction format. But children need to know that the decision is theirs as they too often conform to what they perceive as teachers' expectations. In the calculation $\frac{3}{4} - \frac{1}{5}$, children may be happier with the alternative representation as $0.75 - 0.2$ since both fractions will be familiar as decimals but they may be reluctant to use this knowledge if they think they are meant to work with fractions. This calculation will then depend on subtracting '2 tenths' from '7 tenths and 5 hundredths' and the result 0.55 may be changed back to the fraction for $\frac{55}{100}$ or $\frac{11}{20}$ if required. Converting fractions to decimals can also help to explain the ordering of fractions. A national test question from the Key Stage 2 national assessments in 2002 asked

Which is larger, $\frac{1}{3}$ or $\frac{2}{5}$?

and then

Explain how you know.

Less than half of children working at level 5 and very few children working at level 4 were able to give a successful explanation but one successful strategy was to convert each to a decimal: '$\frac{1}{3}$ is 0.33 (rounded to two decimal places) and $\frac{2}{5}$ is 0.4 which is larger' (QCA, 2003a). It is this flexibility to use alternative representations that is the key to understanding and operating not only with fractions but also with decimals and percentages.

CHAPTER 8

Problem Solving with Numbers

A generation ago there was no question that the written algorithms for 'the four rules' of addition, subtraction, multiplication and division were central to the curriculum and represented the pinnacle of achievement for primary school mathematics. In the curriculum today, however, careful calculating is not enough and there is more emphasis on problem solving as the key to developing mathematical thinking.

Problem solving comes within the 'Using and Applying' strand of the Primary Framework where there are five specific themes:

- Solving problems
- Representing – analyse, record, do, check, confirm
- Enquiry – plan, decide, organise, interpret, reason, justify
- Reasoning – create, deduce, apply, explore, predict, hypothesise, test
- Communicating – explain methods and solutions, choices, decisions, reasoning.

(DfES, 2006b)

From practical problems in the Foundation Stage and Key Stage 1, explaining choices and decisions, to communicating mathematically using precise mathematical language in Key Stages 2 and 3 there is a progression in the development of skills, and children have to be given time and space to tackle problems in mathematics lessons. Guidance says:

> devoting specific lessons to problem solving will help but embedding them into everyday lessons will provide the frequent and regular practice and consolidation children need. Problem solving should not be seen as a 'Friday-only' activity.
>
> (DfES, 2006a)

All the necessary calculating operations can be learned through problem solving, addressing new skills where they are needed, rather than practising isolated calculating procedures. When children solve problems they can use numbers meaningfully, reasoning about their calculations, represent their ideas using words and symbols, and communicate their thinking both orally and with written results. Working with numbers found in contexts, or in patterns and puzzles, will provide a more 'natural' introduction for arithmetic operations so that they are not detached from the reality of the children's lives.

The importance of problem solving in the Foundation Stage is discussed by Gifford (2005) who notes among its benefits motivating learning and making new connections with existing knowledge, 'it involves all the major cognitive learning processes, in visualising solutions, checking for errors, and in a collaborative context, imitation and instruction as well as talking and reflecting' (Gifford, 2005: 151).

These benefits can persist throughout primary school but problem solving can present a challenge in the classroom for the teacher. In her article 'The problem with problem solving' Lesley Jones identifies difficulties including the fact that when children are given freedom to tackle problems they can choose different routes and the teacher's 'control' of learning objectives and outcomes is not so clearly identified. But it is within problem solving that calculations are used purposefully. 'A mathematics curriculum without problem solving can be likened to a diet of PE in which children practise football and netball skills but never get to play a game' (Jones, 2003).

In this chapter the use of number sense in problem solving will be discussed and particular examples will be taken from national tests to illustrate the flexibility in thinking that is needed in solving them. Progression will highlight increasing complexities from one-step to multi-step problems and problems where less routine approaches are needed to solve them. What will typify the problems considered in this chapter is the way few can be solved by direct application of standard written methods for calculating and all require number sense.

Missing numbers and missing operations

The emphasis in recent years has been for assessment questions to test understanding of the operations as well as computational skills. In Chapter 4 the role of questions with missing numbers was discussed but it is also possible to make the operation the missing part. This is shown in the Key Stage 1 question from the 2001 national assessments:

> Write the correct sign in each box.
> 58 □ 26 = 84
> 43 □ 17 = 26
> 33 □ 33 = 0

Although the calculations involved are quite complex for Key Stage 1, a 'feel' for the numbers can be the key to identifying the operations involved as addition and subtraction. It is probable that many children who could not complete this task could have responded thoughtfully to the questions 'Could it be addition?' 'Could it be subtraction/multiplication/division?' This type of thinking is the essence of problem solving – before calculating identify what is needed and what would be a good strategy for solving the problem. It is the difference between 'giving' children a solution strategy and getting them to think for themselves what is the best way for calculating.

In some questions calculating is inevitable but the format of the question can be quite different from the calculating procedures practised in class. In the same test another number problem assessed understanding of the '=' sign as well as the ability to calculate:

> 60 − 40 = 20 + □

The answer zero would have caused concern for children who had never before experienced zero as the answer to a calculation.

A question at Key Stage 2 in 2002 that similarly assessed understanding of the '=' sign was more open-ended and there were many possible solutions:

> Write in what the missing numbers could be:
> 170 + □ = 220 − □

From the responses of the children it was clear that many treated this like the straightforward calculation 170 + □ = 220 and gave the answer 50 for the first missing number. The second box was then filled with some random number, presumably so that the question was completed.

There are so many different possibilities for constructing 'empty box' questions that it is not sensible for children to be taught distinct strategies for coping with each type. They need to make sense of each question focusing on the way different operations and way numbers are connected. Experiences with making up their own questions with missing numbers and missing operations, or swapping questions with a partner, can provide a creative task with discussion focused on the number sense required for a solution.

Calculation may not be necessary if the problem can be related to a known fact as in the problem

$$633 - \square = 34$$

which appeared in the Key Stage 2 national tests in 2006. It seems that individuals make their own sense of the calculation and 667 was a common response (QCA, 2006a).

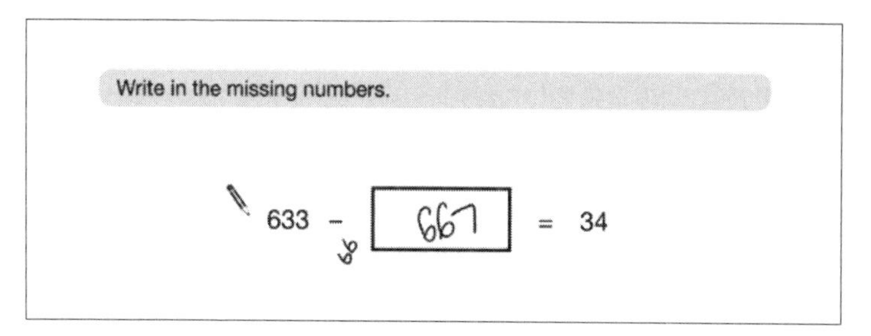

Figure 8.1 An attempt to find the missing number

This is a good example of a calculation where using number sense is so much more important than having standard methods for addition and subtraction. This particular 'missing number' question can be solved mentally if it is related to 633 – 600 = 33 which would provide an effective strategy for its solution. Although not a missing number calculation, another example from the Key Stage 2 tests in 2004 could be solved easily by making a connection, but this time involving division and multiplication: $900 \div (45 \times 4)$. Pupils who recognized $45 \times 2 = 90$ and understood the associative rule could have found the answer 5 very quickly. It is not necessary to know the name for this rule but to have implicit understanding so that the calculation could be thought of as $900 \div (90 \times 2)$ which would be half of $900 \div 90$. This shows the importance of understanding the effects of halving and doubling and using the associative rule discussed in Chapter 5. When an adult sees this question there is probably a strong signal of a connection between 900 and 45 (or 90 and 45) and this type of 'trigger' is crucial in identifying strategies for minimizing the calculating. This is characteristic in good mathematical thinking and can be the objective in teaching number sense. As discussed in previous chapters, this idea of deriving new information from some given fact is a good way for children to explore calculations and develop a 'feel for numbers'. What is important in such questions is the connections that can be made between numbers and operations and these connections need to be the focus for extensive investigative activities and discussion.

Problems involving multi-step calculations

It is not only the complexity of a calculation that can make a problem difficult but some problems involve more than one step in their solution. Typical is the type of question illustrated in Chapter 6 involving money where a total needs to be found and then the change from a given sum calculated. The following problem from the 2003 Key Stage 3 national tests caused considerable difficulty:

> Alice and Ben each buy a bicycle but they pay in different ways.
> Alice pays £179.99.
> Ben pays £8.62 every week for 24 weeks.
> Ben pays more than Alice.
> How much more?
> Show your working.

'Almost a quarter of pupils omitted this question, which suggests that they were unable to analyse the problem and decide on the first step to take' (QCA, 2004b). Just over a tenth of pupils were awarded two marks with about another fifth gaining one mark. This shows the importance of identifying all the steps that are needed to solve a problem and teachers need to help their pupils in planning a strategy. The first step is to extract the relevant information in order to do the calculation and Mike Askew writes about 'uncoupling' from the real-world context and moving around the mathematical world instead (Askew, 2003). In his chapter on 'word problems' he proposes that there are opportunities for 'rich mathematical discussions' where the processes involved in finding solutions are made explicit. He uses the notion of 'horizontal' mathematization to describe the act of setting up the mathematical representation or model for the problem, and 'vertical mathematization' for working with the symbols. These ideas taken from *Realistic Mathematics Education* (Treffers, 1991) help identify the strategies involved in problem solving in a way that can be generalized for all similar problems. This discussion of strategies for problem solving is another example of conceptual discourse in the classroom where the analysis of processes becomes the focus for discussion rather than one particular problem.

This need to break down the approaches needed for multi-step questions begins in the earliest years of schooling as the following Key Stage 1 question from 2003 illustrates:

> There are **60** sweets in a bag.
> **20** are red.
> **16** are yellow.
> The rest are green.
> How many sweets are green?

The common error here was to find the total of the red and yellow sweets but then fail to find the difference between this total and 60 (QCA, 2004a). Children who are used to practising one-step calculations may be aware of the necessary calculating but be inhibited by the difficulty in recording their findings. In this case it is not only strategic planning 'what do I need to do?' that is important but also 'how am I going to write it down?' Children need to be encouraged to write down their strategies and for some this may require a substantial effort. Some children are happy to use jottings but others see such informal working as inappropriate, particularly for a test even though their use has been endorsed as effective in national test reports. The reluctance to write solutions down in their own way may be particularly associated with teaching where there has been great emphasis on standardizing written procedures for the four operations.

When problems are taken from everyday situations or from puzzles where there is no standard approach the focus for teaching will need to include appropriate ways to record the steps that will lead to a solution. An example at Key Stage 1 could be 'There are 14 girls and 12 boys in the class. If they are put in teams of 4 how many teams will there be?' In this case there are many ways to tackle the calculation and record a solution. At Key Stage 2 planning activities can involve complex multi-step calculations. Giving small groups of pupils a catalogue each and a proposed £1000 to spend, what would they buy? Some informal recording would be essential for this activity and the teacher could ask for some record to verify the 'shopping'. A puzzle called 'Differs' comes from the nrich website (www.nrich.mathematics.org.uk) and is more structured but again requires some consideration for recording:

> Choose any 4 whole numbers, for example (100, 2, 37, 59). Now take the difference between consecutive numbers, always subtracting the smaller from the larger, and ending with the difference between the first and the last numbers. What happens when you repeat this process over and over again? For example, continue the sequence of 4-tuples:
> (100, 2, 37, 59), (98, 35, 22, 41), (63, 13, 19, 57), (50, 6, 38, 6), . . .
> Investigate starting with different sets of 4 numbers.

As a mental activity the recording for this activity is easy and children may be encouraged to use jottings to help with the calculating.

Even an apparently straightforward calculation such as the Key Stage 2 question from the 2006 national assessments can involve multiple steps:

> Calculate $\frac{3}{4}$ of £15

Conversion to decimals will result in the single calculation 0.75×15 with a calculator but using fractions requires 15 to first be divided by 4 and then multiplied by 3 (or first multiplied by 3 and then divided by 4) and these steps need to be done consecutively.

In other questions the calculation must be done in separate steps as the following example from the Key Stage 2 national assessments in 2004 shows:

$$100 - (22.75 + 19.08) = \square$$

This calculation is intended to assess children's ability to use a calculator, entering the operations in the correct order and perhaps using the memory key for a partial total. Many children will not be confident in using the memory key and an effective method will be to combine a jotting for the partial total 22.75 + 19.08 worked out on the calculator, followed by clearing the display and calculating independently 100 – 41.83.

Alternatively the partial total 41.83 can be put into the memory by pressing the 'M+' key. Then key in '100', '–', 'MR', '=' which will subtract the number held in memory (MR stands for memory recall) from 100.

Guidance on solving problems involving multi-step calculations states that:

> Children need to be familiar with the order of operations so that they choose the correct sequence in calculations that involve more than one step. They also need to practise jotting down parts of a calculation as they go along. Calculations such as: 8 × (37 + 58), 43 per cent of £285, or ⅜ of 980 km are all multi-step and need to be taught and practised. At Key Stage 2 there should be no particular need to use a calculator to work out percentages, but if necessary the percentage can be represented as a fraction or decimal. So 43 per cent of £285 can be calculated as 285 [×] 0.43, or as 285 [×] 43 [÷] 100. Basic calculators usually have a memory. While there is no requirement for children to use the memory at Key Stage 2, children in Year 6 will probably enjoy learning to use it.
>
> (DfES, 2006d)

Problems requiring logical thinking

All calculating requires logical thinking and making connections but this is more explicit in certain types of number problems. Where there are multiple decisions to be made each with different consequences the use of jottings can help support a systematic approach. The following question from the Key Stage 2 assessments in 2006 is such an example (Figure 8.2).

Each missing digit in this sum is a **9** or a **1**

Write in the missing digits.

☐☐ + ☐☐ + ☐☐ = 201

Figure 8.2 A missing number question from KS2 2006

The logical thinking here starts with the units because there is only one way to combine three numbers, all either 1 or 9, to get 1 as the unit digit in the total 201. Then the overall size of the total 201 would suggest two numbers in the nineties and one with 1 as the first digit. This requires strategic thinking rather than extensive trials which would be very time consuming. Such strategic thinking develops out of discussion rather than calculating practice and cannot be left for children to discover on their own. Because some children develop this type of strategic thinking more intuitively than others there will need to be a balance between shared thinking among pupils and carefully orchestrated opportunities for all pupils to think for themselves. In some cases a 'problem for the week', with strategies discussed only after an extended period, can provide this opportunity.

In both Key Stage 1 and key stage 2 there have been questions that require the ability to make choices and cope with more than one criterion or constraint at the same time. Such an example is the following from the Key Stage 1 national assessment in 2003. Having asked in the first part of the question for the largest number that could be made using each card once, the second part of the question involved two constraints (Figure 8.3).

8	9	7

Use each card **once** to make the smallest **even** number.

Figure 8.3 A question from KS1 2003

Many children made the smallest number but failed to implement the second criterion and make it even. Once again strategic thinking is needed to identify 8 as the only card for the unit position in making an even number and then 798 as the smallest even number.

A similar question at Key Stage 2 from 2002 involved 4 numbers (Figure 8.4).

3	8	9	1

Choose three of these number cards to make an **even** number that is greater than **400**.

Figure 8.4 A question from KS2 in 2002

Once again, where children were unsuccessful with this question it was generally because they wrote a number that satisfied only one of the criteria – for example a number greater than 400 that was not even. Although some terms in the question are given in bold to help identify key elements it can be helpful for children to additionally underline for themselves the key words in the question to help identify all the criteria.

At Key Stage 3 there are questions that are similar but may require more complex decisions and generally involve larger numbers (Figure 8.5).

This example from 2003 may look like an open-ended task that could engage children in investigating connections among numbers. Closer inspection will reveal that constraints have been imposed that limit the number of possible solutions. For the child

Given six number cards

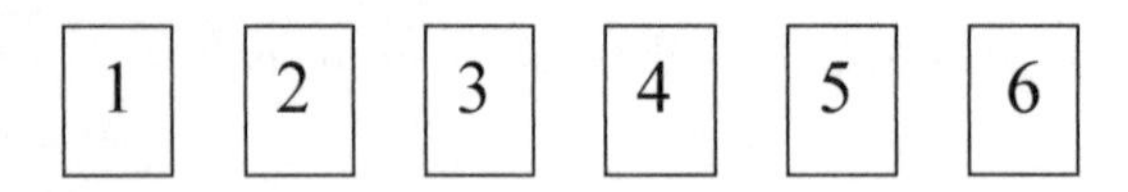

Arrange the six cards to make the given totals. The first one is given

948 = 432 + 516

1164 = □□□ + □□□

750 = □□□ + □□□

Now make a **difference** of 115

115 = □□□ − □□□

Figure 8.5 A question from KS3 in 2003

who has spent extensive time practising standard calculation methods this may prove to be a daunting question as it needs a more flexible and thoughtful approach. It is again number sense that suggests it may be good to look first at the unit entries for each of the calculations. With only the numbers 1, 2, 3, 4, 5 and 6, the total 1164 can only be made using addition by combining 3 and 1 in the units position to get 4 and it does not matter which way round they go. Number sense again suggests that the next place to look is at the overall size of the total, more than 1000. It suggests big numbers in the hundreds place for each number and putting 5 and 6 here quickly leads to a solution 543 + 621. This way of thinking logically about numbers comes from experiences that can be gained through class discussion of these types of problems and through games and puzzles like those found on the nrich website (www.nrich.mathematics.org.uk).

When considering the subtraction part, to make the difference of 115, it is not difficult to see that this introduces complexities greater than addition by the very nature of the operation. In the first place the lack of commutativity means there is less flexibility in placing the units even where they are identified as 6 and 1. This particular question is not as difficult as it could be as no 'borrowing' is needed in order to construct the solution 536 − 421. What if the answer had been given as 295? In this case the remaining digits, 2, 3, 4 and 5, are more difficult to place and some element of 'trial and improvement' may be needed together with some jottings to keep track. The method of 'trial and improvement' can be fruitful if it is used in conjunction with careful logical thinking. To arrive at 9 in the solution the digits must differ by 1 with the larger taken from the smaller. This is the type of complex thinking that can arise from discussion of subtraction calculations rather than just practice. But this is in line with the central message found in national guidelines for teaching mathematics as the following quotes indicate: 'The purpose of teaching algorithms for calculating is to help pupils understand number and use it effectively. Written algorithms are only one possibility' (Learning and Teaching Scotland, 1991).

In the guidelines for England this call for understanding is not so well documented although it is found in the detailed guidance: In Year 3 pupils 'Develop and use written methods to record, support or explain addition and subtraction of two-digit and three-digit numbers' (DfES, 2006c). Overall the English guidance stresses efficiency in calculating which will not facilitate the deeper understanding that is needed for the problem types that typify national tests in England. Rather than practising many calculations, insights explicitly related to this type of thinking will be developed through discussion with children about one calculation or a few related examples. When they are asked to identify the relative difficulty of each example and the different solution strategies that are possible they develop awareness of different characteristics that will influence their approaches to each problem. Once again this will typify the conceptual discourse that supports children's learning (Anghileri, 2006a).

Most of the questions above involve addition and subtraction but there are many similar questions involving multiplication and division that also require strategic thinking. Again, a frequently occurring type of assessment problem involves the arrangement of given cards to make a valid calculation as illustrated in the following national test question from the 2004 Key Stage 2 national assessments (Figure 8.6).

Here are five digit cards.

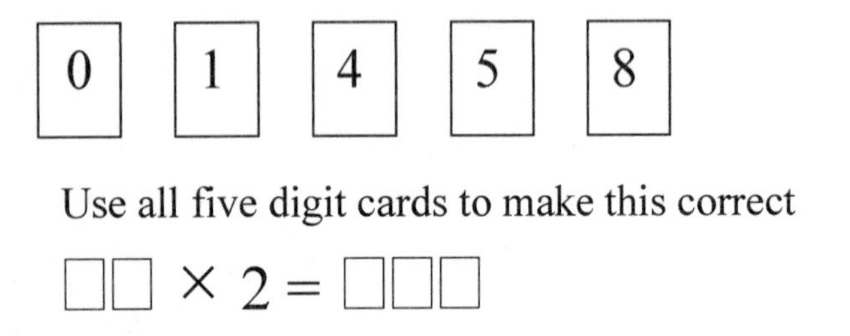

Figure 8.6 A question from KS2 in 2004

For this example number sense must be used to work out logically some constraints as there are 120 possible choices for positioning the cards. Rather than looking for the unit digits this time, the overall size of the three-digit product suggests the first digit card must be 5 or 8 so that a three-digit product is made. The idea of multiplication by 2 invites the remaining digit 4 to be doubled and give the final digit in the product as 8. This leads quickly to $54 \times 2 = 108$. Another approach involves ruling out positions that are not possible. Looking at the overall size is still a good start but then noting the second digit card cannot be 0 or 1 or 8 again suggests 54 as the only possibility for the first number (84 is ruled out as 8 is needed for the final digit in the product).

A subtle difference is needed in the number sense required for this similar question from the 2004 Key Stage 2 national assessments:

> Use the digits **2, 3** and **4** once to make the multiplication that has the **greatest product.**
>
> □□ × □

And the calculations $42 \times 3 = 126$ and $32 \times 4 = 128$ give very close products in a way that may not be easily predicted. The logical argument is that the two biggest numbers, 3 and 4, must be used for the multiplier and the tens part of the other number, and then trialling leads to the result. These two examples together show that number sense must be used flexibly and that experiences with this type of problem through every Key Stage are essential.

One final example is even more open ended. This shows a level 5 question from the 2003 national assessments at Key Stage 2 correctly completed by over half level 5 children, showing the substantial thinking that is achieved by some children (Figure 8.7).

Debbie has a pack of cards numbered from 1 to 20

She picks four different number cards

?	?	?	?

Exactly three of the four cards are multiples of 5

Exactly three of the four numbers are even numbers

All four of the numbers add up to less than 40

Write what the numbers could be.

Figure 8.7 A question from KS2 in 2003

Before embarking on such a problem it is worthwhile planning a strategy and this can be the result of many discussions in class about not only *how to do the calculation* (in the case above a calculator was available) but also *how to record the working* in order to keep track of the thinking involved.

Problems leading to algebra

Algebra has always held a central position in mathematics teaching because of the formalizing and abstracting that it introduces, which are both essential characteristics of well developed mathematical thinking. It takes the focus away from calculating to the identification and implementation of patterns and rules that characterize calculating. Algebra is often associated with the use of letters instead of numbers and the manipulation of equations and formulae according to specified rules. Although this may be relevant to some examples there are many aspects of algebra that involve rules for making patterns and involve no letters. The Framework for Teaching Mathematics states that 'formal algebra does not begin until Key Stage 3 but you need to lay the foundations in Key Stages 1 and 2 by providing early algebraic activities from which later work in algebra will develop' (DfEE, 1999). This includes

- 'forming equations' using symbolic forms such as $16 + 8 = 24$ to express numerical relationships;
- 'solving equations' such as $\square + \Delta = 17$, or inequalities like $1 < \square < 6$;
- 'using inverses' deriving subtraction facts from related addition facts, and using a multiplication fact such as $9 \times 6 = 54$ to derive quickly a corresponding division fact: $54 \div 6 = 9$;
- 'finding equivalent forms' emphasising from the very beginning the different ways of recording for example $24 = 20 + 4 = 30 - 6$ or $30 = 6 \times 5 = 3 \times 2 \times 5$

And also 'identifying number patterns', 'expressing relationships', 'factorising numbers' and 'understanding the commutative, associative and distributive laws' (DfEE, 1999: 10).

The following example from the 2002 Key Stage 2 national assessments involves reversing the rules using the idea of inverses for a calculation where the answer is given:

> Riaz thinks of a number.
> He says
> 'Halve my number and then add 17
> The answer is 23'
> What is Riaz's number?

In this question the order in which the operations are implemented is crucial. The reverse of 'halve and then add three' would be 'subtract 3 and then double' as illustrated in Figure 8.8.

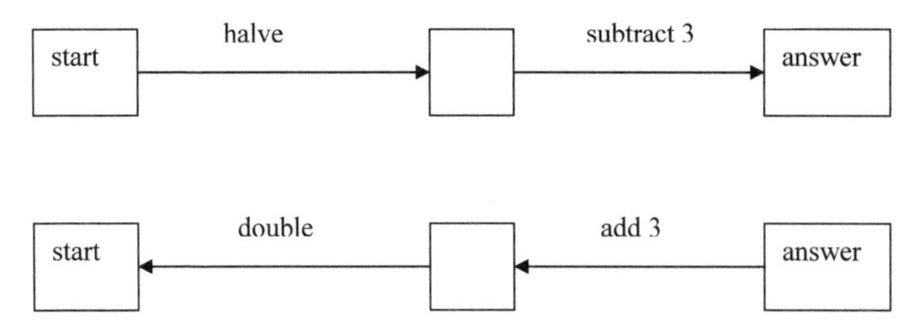

Figure 8.8 Diagram showing inverse operations

This question assesses understanding of inverse operations and the actual calculations are very easy. Although some teachers will not recognize such a question as 'algebra' as it does not involve any letters, it is typical of the early stages of algebraic thinking and fits with the Framework (DfES, 2006a) core learning about inverse operations.

Another non-standard question from the Key Stage 2 assessments of 2003 again gives the result of a calculation and requires the 'starting number':

> Three-quarters of a number is 48
> What is the number?

In this question an approach using number sense would be to find one-quarter from the given three-quarters. This could be signalled by reading the question aloud and relies on an understanding that three-quarters is the same as 3 quarters.

Another way in which algebra thinking is involved is in the construction of a sequence of numbers according to a specified rule which comes in the 'using and applying' mathematics strand of the Primary Framework (DfES, 2006a). This states that in Year 6 'most children will . . . represent and interpret sequences, patterns and relationships involving numbers and shapes' and 'construct sequences and describe the general term'. Children will already be used to sequences as number patterns where each number is derived from the previous one adding a particular number, for example 27, 37, 47, . . . or by multiplication: 3, 6, 9, . . . Other 'rules' may be used, for example starting at 1 and adding 3 will result in 1, 4, 7, 10, . . . or starting at 10 and subtracting 3 each time to give 10, 7, 4, 1, −2, −5, . . . More complex sequences can be constructed, for example start at 1, double each number and add 3 which will result in the sequence 1, 5, 13, 29 . . . 'Guess my rule' is a game that can be played with a class where either the teacher or a pupil shows a sequence of numbers made according to a rule that the rest of the class have to work out. Sequences appear frequently in national test questions as the example (Figure 8.9) from Key Stage 2 national assessments for 2006 shows.

> The numbers in this sequence increase by the same amount each time.
>
> Write in the missing numbers

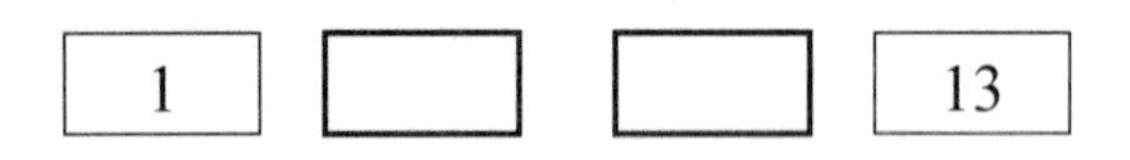

Figure 8.9 A question from KS2 in 2006

In finding a solution to this question an empty number line could be helpful. The question then becomes one of finding two equal jumps to get from 1 to 13. The empty number line will help illustrate all sequences where the numbers differ by the same amount. This type of sequence is called an *arithmetic progression* and can be identified by finding the same *common difference* between each of the consecutive terms. When each term is obtained by multiplying (or dividing) each number by a constant term (the *common ratio*) to get the next number this sequence is called a *geometric progression.*

When the rule for construction is given the sequence can be extended indefinitely. Another sequence example from Key Stage 2 national assessments in 2004 gives the rule for construction but asks for a number so far away from those that are given that there is no possibility of finding all the interim numbers. In this example the rule needs to be reversed, again with careful attention to the order:

> A sequence of numbers starts at 11 and follows the rule 'double the last number and then subtract 3'

> 11 19 35 67 131 ...
> The sequence continues.
> The number 4099 is in the sequence.
> Calculate the number immediately **before 4099** in the sequence

This question combines understanding of the rule for constructing the sequence with understanding of inverse operations so that 'double the last number and subtract 3' is reversed to give '4099 add 3 and then halved'.

Letters instead of numbers

All of the previous examples require algebraic thinking but the type of problems most teachers associate with algebra are those which involve letters to represent numbers. The Framework (DfES, 2006a) states that in 'Year 6 progression to Year 7 most children learn to ... generate sequences and describe the general term: use letters and symbols to represent unknown numbers or variables'. The transition from arithmetic to algebra is problematic because children can work intuitively with problem solving in arithmetic but need to understand the 'rules' for working with symbols. Nickson talks of the 'jump children are expected to make from the "reality" of arithmetic to the more abstract world of algebra' (Nickson, 2004). From the earliest stages of recording, a box will have been used to represent a missing number but the introduction of a letter appears to confuse many children (and adults). Rewriting $\square + 85 = 200$ as $x + 85 = 200$ the unknown number is given the name x but for many adults this will be associated with a graph or other mathematical ideas. More complex would be the problem $170 + \square = 220 - \Delta$ re-written as $170 + x = 220 - y$ because the unknowns can take on more than one value and because the introduction of two variables will later be identified with line graphs. At Key Stage 2 none of the further implications are known and the letter can be treated in exactly the same way as the box or the triangle. Procedures will be the same as those for finding the missing numbers discussed above.

Making sense of the information given and testing suggestions will involve jottings to keep track of trials and improve the suggestions. The following question from the 2006 Key Stage 2 national assessments is such an example:

> k stands for a whole number.
> k+7 is greater than 100
> k − 7 is less than 90
> Find **all** the numbers that **k** could be.

In this example the letter k can be treated in the same way that a box is used in the missing number problems above but this time the box could be filled with several numbers instead of a unique one. Tackling this problem relies mostly on an individual's confidence to interpret the question and then to use number sense. Children will gain this confidence by working together with the teacher to build experience in reading aloud and interpreting this type of question. 'Some number, when I add 7 to it gets bigger than 100. What could that number be?' This will result in a list of possible values of k that fit the first constraint. Similarly 'some number when I subtract 7 from it will be smaller than 90. What could this number be?' will result in another list. Satisfying both constraints at the same time requires the numbers that are in both lists.

In teaching mathematics it is no longer appropriate to show children what to do and get them practising a repertoire of standard procedures for calculating. The examples in this chapter and throughout the book show that teachers today need to work together with children to develop the thinking and understanding of individuals who will become the problem solvers of the future. We need to give them confidence to tackle novel situations with the satisfaction of success and the desire to go further. While we are all still striving to establish how to achieve this it makes teaching mathematics an intellectual challenge that will present many frustrations but immense satisfaction as children realize the power of their own mathematics.

References

Anghileri, J. (1995) 'Language, arithmetic and the negotiation of meaning', *For the Learning of Mathematics* 21(3), 10 [reprinted in QCA '*National Numeracy Strategy: Guide for your professional development*: Book 3'].

Anghileri, J. (2001) 'Intuitive approaches, mental strategies and standard algorithms', in J. Anghileri (ed.) *Principles and Practices in Arithmetic Teaching*. Buckingham: Open University Press.

Anghileri, J. (2006a) 'Scaffolding practices that enhance mathematics learning' *Journal of Mathematics Teacher Education* 9(1), 33–52.

Anghileri, J. (2006b) *Teaching Number Sense* (second ed.). London: Continuum.

Anghileri, J. and Beishuizen, M. (1998) 'Counting, chunking and the division algorithm', *Mathematics in School* 27(1), 2–4.

Anghileri, J., Beishuizen, M. and Van Putten, C. (2002) 'From informal strategies to structured procedures: mind the gap!', *Educational Studies in Mathematics* 49(2), 149–70.

Askew, M. (2001) 'What does it mean to learn? What is effective teaching?', in J. Anghileri, J. (ed.) *Principles and Practices in Arithmetic Teaching*. Buckingham: Open University Press.

Askew, M. (2003) 'Word problems: Cinderellas or wicked witches?', in Ian Thompson (ed.) *Enhancing Primary Mathematics Teaching*. Maidenhead: Open University Press.

Askew, M., Brown, M., Rhodes, V., Wiliam, D., and Johnson, D. (1997) *Effective Teachers of Numeracy: Report of a study carried out for the Teacher Training Agency*. London: King's College, University of London.

Beishuizen, M. (1999) 'The empty number line as a new model', in Ian Thompson (ed.) *Issues In Teaching Numeracy In Primary Schools*. Buckingham: Open University Press.

Beishuizen, M. (2001) 'Different approaches to mastering mental calculation strategies', in J. Anghileri (ed.) *Principles and Practices in Arithmetic Teaching*. Buckingham: Open University Press.

Beishuizen, M. and Anghileri, J. (1998) 'Which mental strategies in the early number curriculum?', *British Education Research Journal* 24(5), 519–38.

Brown, M. (1982) 'Place value and decimals', in Hart, K. (ed.) *Children's understanding of mathematics 11–16.* London: Murray.

Buys, K. (2001) 'Progressive mathematization: a sketch of a learning strand', in Anghileri, J. (ed.) *Principles and Practices in Arithmetic Teaching.* Buckingham: Open University Press.

Cockcroft, W. (1982) *Mathematics Counts: Inquiry into the Teaching of Mathematics in Schools.* London: HMSO.

DfEE (Department for Education and Employment) (1999) *The National Numeracy Curriculum: Handbook for primary teachers in England Key Stages 1 and 2.* London: DfEE.

DfEE (Department for Education and Employment) (2001) *The National Numeracy Strategy Framework for Teaching Mathematics: Years 7, 8 and 9.* London: DfEE.

Fischbein, E., Deri, M., Nello, M. and Marino, M. (1985) 'The role of implicit models in solving verbal problems in multiplication and division', *Journal for Research in Mathematics Education* 16, 3–17.

Forrester, R. (2003) 'It's not calculators but how they're used ....', in Way, J, and Beardon, T. *ICT and Primary Mathematics.* Maidenhead: Open University Press.

Foxman, D. and Beishuizen, M. (1999) 'Untaught mental calculation methods used by 11-year olds: some evidence from the APU Survey in 1987', *Mathematics in School,* 28(5).

Gifford, S. (2005) *Teaching Mathematics 3–5.* Maidenhead: Open University Press.

Hart, K. (ed.) (1982) *Children's Understanding Of Mathematics 11–16.* Windsor: NFER-Nelson.

Heirdsfield, A. (2005) 'One teacher's role in promoting understanding in mental computation', in Chick, H. and Vincent, J. (eds) *Proceedings of the Twenty-ninth International Conference for Psychology in Mathematics Education* vol. 3, 113–20. University of Melbourne.

Hemmings, F. (1932) *The Teaching of Arithmetic and Elementary Mathematics.* London and Glasgow: Blackie and Son Ltd.

Jaworski, B. (1988) '"Is" versus "seeing as": constructivism and the mathematics classroom', in D. Pimm (ed.) *Mathematics, Teachers and Children.* London: Hodder and Stoughton.

Jones, L. (2003) 'The problem with problem solving', in Ian Thompson (ed.) *Enhancing Primary Mathematics Teaching.* Open University Press: Maidenhead.

Menne, J. (2001) 'Jumping ahead: an innovative teaching programme', in J. Anghileri (ed.) *Principles and Practices in Arithmetic Teaching.* Buckingham: Open University Press.

Moss, J. and Case, R. (1999) 'Developing children's understanding of the rational numbers: a new model and an experimental curriculum' *Journal for Research in Mathematics Education* 30(2), 122–47.

National Curriculum online www.nc.uk.net

Nickson, M. (2004) *Teaching and Learning Mathematics: A teachers' guide to recent research and its application* (second ed.). London: Continuum.

Perks, P. and Prestage, S. (2003) 'Spreadsheets with everything', in Way, J. and Beardon, T. *ICT and Primary Mathematics.* Maidenhead: Open University Press.

Pimm, D. (1987) *Speaking Mathematically: Communication in Mathematics Classrooms.* London: Routledge and Kegan Paul.

QCA (Qualifications and Curriculum Authority) (1999) *Teaching Mental Calculation Strategies.* London: QCA.

QCA (Qualifications and Curriculum Authority) (2002) Standards at Key Stage 2 – a report for headteachers, class teachers and assessment coordinators on the 2001 national curriculum assessments for 11-year-olds. London: QCA

QCA (Qualifications and Curriculum Authority) (2003a) *Standards at key stage 2 – English, mathematics and science: a report for headteachers, class teachers and assessment coordinators on the 2002 national curriculum assessments for 11-year-olds.* London: QCA.

QCA (Qualifications and Curriculum Authority) (2003b) Standards at Key Stage 1 – English and Mathematics: a report for headteachers, class teachers and assessment coordinators on the 2002 national curriculum assessments for 7-year-olds. London: QCA.

QCA (Qualifications and Curriculum Authority) (2004a) *Standards at key stage 1 – English and mathematics: a report for headteachers, class teachers and assessment coordinators on the 2003 national curriculum assessments.* London: QCA.

QCA (Qualifications and Curriculum Authority) (2004b) *Standards at key stage 3 – Mathematics: a report for headteachers, heads of department, mathematics teachers and assessment coordinators on the 2003 national curriculum assessments.* London: QCA.

Riley, M., Greeno, J. and Heller, J. (1983) 'Development of children's problem solving ability in arithmetic', in Ginsburg, H. (ed.) *The Development of Mathematical Thinking.* New York: Academic Press.

Rousham, L. (2004) 'CAN calculators make a difference', in Julia Anghileri (ed.) *Children's Mathematical Thinking in the Primary Years.* London: Continuum.

Schonell, F. (1937) *Diagnosis of Individual Difficulties in Arithmetic.* Edinburgh and London: Oliver and Boyd.

Smith, A. (2004) *Making Mathematics Count, the Report of Professor Adrian Smiths Inquiry into Post-14 Mathematics Education.* London: The Stationery Office.

Thompson, I. (1997) 'Mental and written algorithms: can the gap be bridged?', in Ian Thompson (ed.) *Teaching and Learning Early Number.* Buckingham: Open University Press.

Thompson, I. (1999) (ed.) *Issues In Teaching Numeracy In Primary Schools.* Buckingham: Open University Press.

Treffers, A. (1991) 'Didactical background of a mathematics program for primary education', in L. Streefland (ed.) *Realistic Mathematics Education in Primary School.* Utrecht: Freudenthal Institute, Utrecht University.

Van den Heuvel-Panhuizen, M. (2001) 'Realistic Mathematics Education in the Netherlands', in J. Anghileri (ed.) *Principles and Practices in Arithmetic Teaching.* Buckingham: Open University Press.

Van den Heuvel-Panhuizen, M., Middleton, J. and Streefland, L. (1995) 'Student generated problems: easy and difficult problems on percentage', *For the Learning of Mathematics* 15, 3.

Way, J. and Beardon, T. (2003) *ICT and Primary Mathematics.* Maidenhead: Open University Press.

Websites

DfES (2006a) Primary Framework for literacy and mathematics www.standards.dfes.gov.uk/primaryframeworks

DfES (2006b) Primary Framework for literacy and mathematics – Guidance papers – using and applying www.standards.dfes.gov.uk/primaryframeworks/library/Mathematics/guidance/resources

DfES (2006c) Primary Framework for literacy and mathematics – Guidance papers – calculations www.standards.dfes.gov.uk/primary frameworks/library/Mathematics/guidance/resources

DfES (2006d) Primary Framework for literacy and mathematics – Guidance papers – the use of calculators in teaching and learning mathematics www.standards.dfes.gov.uk/primaryframeworks/library/Mathematics/guidance/resources

Learning and Teaching Scotland (1991) *5–14 guidelines for primary and S1/S2 curriculum* www.ltscotland.org.uk/5to14/guidelines/index.asp

National Curriculum online www.nc.uk.net

nrich www.nrich.mathematics.org.uk

QCA (2006a) Implications for teaching and learning www.qca.org.uk/downloads/QCA-06-2496-Problem-solving.ppt#307,2, Problem solving

QCA (2006b) Implications for teaching and learning www.qca.org.uk/downloads/QCA-06-2502-Measures.ppt#318,12,Slide 12

Index